COVID-19 Mask Yes? Or No?

It's Everybody's Recommended Solution!!!

Read this book to learn about what the experts already know.

LETS GO PUBLISH

BRIAN W. KELLY

Editor, Brian P. Kelly
Author Brian W. Kelly
Title: The COVID-19 Mask: Yes? Or No?
Subtitle: It's Everybody's Recommended Solution!!!

Referenced Material: *The information in this book has been obtained through personal and third-party observations, interviews, and copious research. Where unique information has been provided or extracted from other sources, those sources are acknowledged within the text of the book itself or at the end of the chapter in the Sources Section. Thus, there are no formal footnotes nor is there a bibliography section. Any picture that does not have a source was taken from various sites on the Internet with no credit attached. If resource owners would like credit in the next printing, please email publisher.*

Published by: LETS GO PUBLISH!
Publisher: Brian P. Kelly
Editor: Brian P. Kelly
P.O Box 621 Wilkes-Barre, PA www.letsgopublish.com

Library of Congress Copyright Information Pending
Book Cover Design by Brian W. Kelly;
Editing by Brian P. Kelly

ISBN Information: The International Standard Book Number (ISBN) is a unique machine-readable identification number, which marks any book unmistakably. The ISBN is the clear standard in the book industry. 159 countries and territories are officially ISBN members. The Official ISBN For this book is also on the outside cover: **978-1-951562-27-4**

The price for this work is: **$12.95 USD**

10 9 8 7 6 5 4 3 2 1

Release Date: September 2020

LETS GO PUBLISH!

Dedication

I dedicate this book to all of the great health care workers and first responders. God bless you and thank you for your great assistance during the crisis.

Thank You All!

Acknowledgments

I appreciate all the help that I have received in putting this book together as well as all of my other 242 other published books.

My printed acknowledgments had become so large that book readers "complained" about going through too many pages to get to page one of the text.

And, so to permit me more flexibility, I put my acknowledgment list online, and it continues to grow. Believe it or not, it once cost about a dollar more to print each book.

Thank you and God bless you all for your help.

Please check out www.letsgopublish.com to read the latest version of my heartfelt acknowledgments updated for this book. FYI, Wily Ky Eyely loves this book and recommends it to all. Click the bottom of the Main menu on the web site!

Thank you all!

Preface

Why did Brian W. Kelly write this book?

Brian W. Kelly wrote this book because he cares about the United States and the other countries suffering today from the Coronavirus. . I am publishing this book because I care.

I hope you enjoy this book and I hope that it inspires you to take the individual actions necessary to help us all during this crisis.

Government should be a helpful tool in solving this deep moral and potentially existential dilemma for our country.

I wish you the best.

Brian P. Kelly, Publisher
Wilkes-Barre, Pennsylvania

Table of Contents:

Dedication v

Chapter 1 Santa Claus's Mask 1

Chapter 2 COVID-19 Masks of Many Flavors 7

Chapter 3 Masks Work: Sometimes Yes; Sometimes No 15

Chapter 4 Reuse COVID Masks??? 19

Chapter 5 Counterfeit Masks 'R Us 27

Chapter 6 Care for Your Mask 33

Chapter 7 Why People Get Upset Over Wearing Masks 43

Chapter 8 The Hydroxychloroquine Saga 49

Chapter 9 The Remedy Across the World! 55

Chapter 10 How to Resume Your Life 63

Other Books by Brian W. Kelly: (amazon.com, and Kindle) 85

About the Author

Brian W. Kelly retired as an Assistant Professor in the Business Information Technology (BIT) program at Marywood University, where he also served as the IBM i and Midrange Systems Technical Advisor to the IT Faculty. Kelly designed, developed, and taught many college and professional courses. He continues as a contributing technical editor to a number of IT industry magazines, including "The Four Hundred" and "Four Hundred Guru," published by IT Jungle.

Kelly is a former IBM Senior Systems Engineer and IBM Mid Atlantic Area Specialist. His specialty was designing applications for customers as well as implementing advanced IBM operating systems and software facilities on their machines.

He has an active information technology consultancy. He is the author of 243 books and numerous technical articles. Kelly has been a frequent speaker at COMMON, IBM conferences, and other technical conferences.

Brian was a candidate for US Congress from Pennsylvania in 2010 and he brings a lot of experience to his writing endeavors.

Brian Kelly knows that the crisis can be solved with the right techniques in the short term. Keep the faith, we'll be on the streets safely soon.

Chapter 1 Santa Claus's Mask

Do you believe in Santa Claus?

Let's begin this book with a chapter that tells a light-hearted story as a take-off on *The Miracle on 34th Street* in case Santa was caught at Macy's without his COVID-19 mask. What could be worse in 2020? The real idea behind this chapter is to lightly introduce the reader to the idea of what a real COVID-19 mask is and that under current regulations, even Santa Claus would be required to wear one.

During this chapter and the next, we gear up for the seriousness of the book and then we take off to the facts we all need to know about a subject that did not exist just half a year ago.

Though there have been sporadic, OK more than sporadic, complaints about face masks, nobody of whom I am aware has complained in recent times at least, that they have seen a person cooperating with the mask edicts across the country by wearing a Santa Claus COVID-19 mask. But we do have secret proof that there are some. Perhaps not a full beard mask. How about you? Looking at what it might look like, some of the male persuasion may not complain anyway if it looked like a mask such as the one above. Nice mask fellas!

Just maybe. Now, folks, the lady on the prior page is not Santa Claus, though he is nestled comfortably in her protective mask. Of course, the family and the neighborhood kids might not think it is an improvement on the caricature of the Jolly Old Elf that comes down the chimney on Christmas Eve. However, if we, adults of all genders choose to admit, it is a beautiful mask, worn beautifully for sure.

The whole picture might be characterized as a beautiful flower garden. It helps capture some nice sentiments for sure. OK let's get back on track here for awhile. OK!

Can we tell that the above gentleman is not masked properly?
How about this next one?

Santa won't give anybody COVID-19 for sure. It's not in his genes. But, he might be mistaken for a mask-less good ole elf that might.

Might Santa be put into Bellevue Hospital if he were to try to get by with his beard alone as a mask at Macy's in 2020? Would Sawyer turn him into the gendarme for a lockup or lockdown?

We can all recall that in *The Miracle on 34th Street* (Macy's Store) when in the company cafeteria, a young employee Alfred (Alvin Greenman) who today would gladly clad a mask for his personal protection, tells Kris Kringle that Sawyer convinced him that he is unstable simply because he is a kind-hearted mask wearer. Whew!

Kris Kringle naturally gets irate and takes his mask off and immediately goes to Sawyer's office to confront him, eventually striking him on the head with an umbrella that looks like a cane. The seemingly powerful Sawyer feigns his pain and suffering from the "tap" so he could have Kris confined to Bellevue Hospital.

Tricked into cooperating, and believing Doris supplied a picture to Sawyer that showed Kris not wearing his Macy's supplied PPE N95 mask, Kris becomes dejected and he then deliberately fails his examination. He is thus recommended for permanent commitment in the Cuckoo Nest named Bellevue Hospital. However, Fred Galey, Susan's mom Doris's potential beau persuades Kris not to give up. Kris agrees to fight for his freedom.

In the subsequent hearing, the question is whether Kris will admit that other than Sawyer's captured picture without his COVID-19 mask, Kris had been a 100% bona fide mask wearer and thus should be released on his own recognizance. A bevy of witnesses testified to the same at the hearing. Kris' defense team also had to prove that he was not delusional about being the jolly old elf (Santa Claus) which was a bit more challenging. They were up to it.

By the way, while at Bellevue, Kris had his beard "mask" tested to see if it qualified as a COVID-19 mask. The Epidemiologists at Bellevue proved that it was actually more effective against all viruses than the highest quality surgical mask—even better than N95. Unfortunately, for the rest of us, since it was a natural beard that had been grown from Kris's face, the Santa Beard mask could not be easily duplicated and strangely, the mask technicians could not determine from which material elements of earth it was actually composed.

During the court Hearing for Kris's release from Bellevue, he was quite cheerful because he had been convinced through notes and cards and letters that people were believing in him. His attorney, Galey was at first at a loss for a defense, especially after Kris admitted on the stand that he believed himself to be Santa Claus. Despite his SC claim, he did show the people in the court gallery that his wits were still with him and they had not taken the day off. That part is coming up.

One of the key parts of the trial was when the district attorney asked Kris, "Where do you live?" and Kris responded, "That's what this hearing will decide," much to the amusement of the court. Kris admitted that just one time he had his mask off when Sawyer saw him, and he hoped the court would permit him to wear his mask at his home and at his continuing job At Macy's.

Then the hearing was interrupted when Post Office personnel all donned in N95 government issued masks dropped bags of Santa mail (Letters to Santa) on the judge's desk and they presented Kris with a brand new mask with Mrs. Claus and Rudolf neatly stitched on it. The movie then ended on a happy note with Kris being released and going back home to prepare for Christmas Eve.

Joe Biden would more than likely tell us all that the lesson of this seasonal story is "Wear Your Mask!" "Come on Man!"

As in all great COVID masking stories, this ending is no different from the one Scientist Anthony Fauci would have written.

"There are three components Fauci would echo to being safety in today's COVID-19 world as follows:

1. Wear a Mask
2. Wash your hands
3. Maintain 6 feet social distancing
"

If you can manage it with these three minor restrictions, you may then enjoy your life.

Chapter 2 COVID-19 Masks of Many Flavors

Mostly everybody wears a mask in public today

Masks and beards are different for sure. People wearing face masks have been very commonplace in the US for about six months now. The masks came about a few months after Christmas 2019 when the Wuhan virus made itself known in America. Nobody could speak with Kris Kringle at the time about whether his recent Christmas Eve's in china had been uneventful.

Some people of course do wear face masks with pleasure while others such as Santa in the prior page wears his mask reluctantly. As the knowing Claus, He knows that his face grown mask works fine and he sees the mask as redundant.

There is one profession these days whose members are not seeming to complain at all—bank robbers. Here we are in the coronavirus-era, and bank robbers appear to be taking advantage of the mask requirements that the rest of us wear for COVID-19 safety.

Many know that since the nation's first-known armed bank robbery in 1866 by the Jesse James gang, criminals have come up with all sorts of disguises to conceal their identities from employees and security mechanisms. Masks have been the most popular and some of the COVID-19 masks of today are designed to mimic the robber masks of yore.

Paula Nenda, who serves as the senior VP for Risk and Cybersecurity Policy at the American Bankers Association told us recently that "We've never allowed people into banks with masks, you know, simply because of the security risk of having unidentified people in the bank. The challenge we've got is that the pandemic is the larger risk right now." Nonetheless the job of bank robber has just gotten easier.

Ironically, while many of those protecting themselves from Coronavirus use the traditional bandana type robber masks like the James Gang, the robbers don't necessarily measure up to expectations.

"I mean, the exact picture of what you would expect a bank robber to wear, and what people are wearing and getting away with it, the FBI says: "You're almost inviting somebody to come in to a bank and attempt a bank robbery."

So, one gimmick used by banks today to make up for people wearing masks, is to have a bank employee—a greeter—at the entrance of the bank to identify each person who comes in. " The bank has the greeter ask each person to remove their mask temporarily, face the camera, and identify themselves. That gives the bank the information on the potential perpetrator they need to nab a future bank robber. That is the best deterrent but the FBI has others. So, now you can't even safely rob a bank even if you wear the best N95 mask.

We all can relate to the bad guys wearing masks in the old movies but I can recall learning about my first and most favorite masked man on the radio every night at 7:30 P.M. Eastern Time He was a good guy though, as good a guy as you could get. It was The Lone Ranger and he did not go anyplace without his trusty companion, Tonto. Tonto did not wear a mask in the show.

There were other good guys over time that have been brought to us by Hollywood. We've all seen both Batman and Robin, Captain America, Spiderman, Zorro, the Phantom, the Green Lantern, and as noted we've got the inimitable "who was that masked man? --The Lone Ranger.

Back then in the robber era, the mask was a symbol or a way to hide. The gun was the robbery instrument. Today the mask is the issue, good or bad. It is the implement in question. People talk about using the mask for protection or for risking the contraction of the Coronavirus disease because they wear or do not wear the mask. .

There is a big controversy on COVID-19 mask studies which spotlights the messiness of science or pseudo-science during a pandemic. The big controversy arises because there is not 100% evidence of the exact value of a mask in protecting one from contracting the COVID-19.

Some think masks definitely help a lot and some think they do not help much. There is not one authoritarian voice with the power to make the declaration but despite this many choose to take sides without the proper authority. Even Doctor Fauci's last word is not seen as the last word by all on the value of masks. Which is it? Where would we go to find a definitive declaration? Nobody actually knows.

On Tuesday, July 14th for example, the CDC, a self-proclaimed authoritative source on all things COVID-19, put out this report for immediate Release. They asked for questions by contacting their Media Relations staff at 404-639-3286. Here is their release in its entirety.

> "Americans are increasingly adopting the use of cloth face masks to slow the spread of COVID-19, and the latest science may convince even more to do so.
>
> In an editorial published today in the Journal of the American Medical Association (JAMA), CDC reviewed the latest science and affirms that cloth face coverings are a critical tool in the fight against COVID-19 that could reduce the spread of the disease, particularly when used universally within communities. There is increasing evidence that cloth face coverings help prevent people who have COVID-19 from spreading the virus to others.
>
> "We are not defenseless against COVID-19," said CDC Director Dr. Robert R. Redfield. "Cloth face coverings are one of the most powerful weapons we have to slow and stop the spread of the virus – particularly when used universally within a community

setting. All Americans have a responsibility to protect themselves, their families, and their communities."

This review included two case studies out today, one from JAMA, showing that adherence to universal masking policies reduced SARS-CoV-2 transmission within a Boston hospital system, and one from CDC's Morbidity and Mortality Weekly Report (MMWR), showing that wearing a mask prevented the spread of infection from two hair stylists to their customers in Missouri.

[Please note that the tests were on SARS-CoV-2, and not COVID]

Additional data in today's MMWR (Morbidity and Mortality Weekly Report (MMWR) showed that immediately after the White House Coronavirus Task Force and CDC advised Americans to wear cloth face coverings when leaving home, the proportion of U.S. adults who chose to do so increased, with 3 in 4 reporting they had adopted the recommendation in a national internet survey.

The results of the Missouri case study provide further evidence on the benefits of wearing a cloth face covering. The investigation focused on two hair stylists — infected with and having symptoms of COVID-19 — whose salon policy followed a local ordinance requiring cloth face coverings for all employees and patrons. The investigators found that none of the stylists' 139 clients or secondary contacts became ill, and all 67 clients who volunteered to be tested showed no sign of infection.

The finding adds to a growing body of evidence that cloth face coverings provide source control – that is, they help prevent the person wearing the mask from spreading COVID-19 to others. The main protection individuals gain from masking occurs when others in their communities also wear face coverings.

COVID-19 prevention in a Missouri hair salon

When two stylists at a Missouri hair salon tested positive for the virus that causes COVID-19, researchers from CoxHealth hospitals, Washington University, the University of Kansas, and the Springfield-Greene County Health Department worked together to trace contacts, investigate the cases, and publish their findings in the MMWR.

One of the stylists developed respiratory symptoms but continued to see clients for eight days. The other, who apparently became infected from her co-worker, also developed respiratory symptoms and continued to see clients for four days.

The salon in which they worked had a policy requiring both stylists and their clients to wear face coverings, consistent with the local government ordinance. Both stylists wore double-layered cloth face coverings or surgical masks when seeing clients. The median appointment time was 15 minutes and ranged from 15 to 45 minutes. More than 98% of clients wore a face covering—47% wore cloth face coverings, 46% wore surgical masks, and about 5% wore N-95 respirators.

When customers were asked whether they had been ill with any respiratory symptoms in the 90 days preceding their appointment, 87 (84%) reported that they had not. None of the interviewed customers developed symptoms of illness. Among 67 (48%) customers who volunteered to be tested, all 67 tested negative for the virus that causes COVID-19. Several family members of one of the stylist's subsequently developed symptoms and received a diagnosis of COVID-19.

Survey: Acceptance of face-mask guidance increased

CDC analyzed data from an internet survey of a national sample of 503 adults during April 7–9 and found that about 62% said they would follow the newly announced recommendations to wear a face mask when outside the home. A repeat survey during May 11-13 showed that the percentage of adults endorsing face mask wearing increased to more than 76%.

The increase was driven largely by a significant jump in approval by white, non-Hispanic adults, from 54% to 75%. Approval among Black, non-Hispanic adults went up from 74% to 82%, and remained stable among Hispanic/Latino adults at 76% and 77%.

There was also a large increase in face-mask approval among respondents in the Midwest, from 44% to 74%. Approval was greatest in the Northeast, going from 77% to 87%.

Resources:

- CDC's Information on Cloth Face Coverings: https://www.cdc.gov/coronavirus/2019-ncov/prevent-getting-sick/diy-cloth-face-coverings.html
- CDC Editorial in JAMA: Brooks JT, Butler JC, Redfield RR. Time for universal masking and prevention of transmission of SARS-CoV-2. JAMA. Published online July 14, 2020. doi:10.1001/jama.2020.13107 https://jamanetwork.com/journals/jama/fullarticle/10.1001/jama.2020.13107external icon
- MMWR Article: No Transmission of Symptomatic SARS-CoV-2 After Significant Exposure With Universal Face Mask Use at a Hair Salon – Springfield, Missouri, May 2020 https://www.cdc.gov/mmwr/volumes/69/wr/mm6928e2.htm?s_cid=mm6928e2_w

MMWR Article: Factors Associated with Cloth Face Coverings Use during the COVID-19 Pandemic — United States, April and May 2020
https://www.cdc.gov/mmwr/volumes/69/wr/mm6928e3.htm?s_cid=mm6928e3_w

And, so, folks, there is evidence that cloth masks work. If there were definitive scientific proof from the US sponsoring authority, the report would have said so. Thus, we can conclude that it looks like cloth maks are helpful in fighting COVID-19 community spread. I don't know that this is definitive authoritative proof. We can't tell from this.

Chapter 3 Masks Work: Sometimes Yes; Sometimes No

Retraction of a Definitive Study

It is a fact that we have seen studies one week indicate definite positive helpful results from masking while sometimes the very next week another study says no dice—the masks don't work.

For example, about a week before I wrote this paragraph, a group of researchers posted a letter that they had sent to the "Proceedings of National Academy of Sciences (PNAS)." The letter requested the retraction of a study published the week before that purportedly showed mask use was the most effective intervention in slowing the spread of COVID-19 in New York City. To be or not to be, that is the masking question.

Though I have not yet seen a response from the PNAS editors to respond to the retraction request, scientists have roundly criticized the study's methodology, and the entire confliction has highlighted the difficulty of "doing science" amid a full-blown pandemic when the world is waiting for a conclusion.

The paper in question, "Identifying airborne transmission as the dominant route for the spread of COVID-19" states, "After April 3, the only difference in the regulatory measures between NYC and the United States lies in face covering on April 17 in NYC."

The group of scientists, many from Stanford and Johns Hopkins universities, took umbrage with that conclusion and said it is verifiably false on several accounts: Other parts of the country had mandated mask use, and different parts of the United States had different degrees of "lockdown."

"NO DEFINITIVE PROOF"

"While masks are almost certainly an effective public health measure for preventing and slowing the spread of SARS-CoV-2, the claims presented in this study are dangerously misleading and lack any basis in evidence." They wrote this in a letter to the PNAS editorial board, requesting retraction. "Unfortunately, since its publication on June 11 2020, this article had already been distributed and shared widely in traditional and social media, where its claims had been interpreted as rigorous science."

But, is it?

Additionally, does slowing the spread of SARS-CoV-2, slow the spread of COVID-19? No comments in this report.

Stakes are much higher than before'

People report on the letter's status. Noah Haber, ScD, a postdoctoral fellow at Stanford University, said he has heard from PNAS editors that they have received the letter. Haber was the first co-signer of the letter requesting retraction.

"The policy implications of this paper is immediate, so we hope that the response is commensurate with the decisions that need to be made," he told CIDRAP News. Either there is proof or there is not? It can't be both ways. What can one conclude when the experts find it too messy to make a conclusion?

Haber said he and his colleagues are not arguing the usefulness of masks. They implicitly believe the masks are good. However, they want the report to be correct. They point out that the study in question cannot be used to prove the conclusions one way or another. It could not evaluate how effective masking policies are relative to other policies.

The problem is that they had released the paper as if it were proof of something. Yet, "there are an enormous number of severe errors with the paper," Haber said. "Unfortunately, this is not a new problem in science, but the stakes are much higher than before." With the community looking for quick results, the science sometimes cannot prove one way or another without a proper amount of time passing.

Haber said the paper also highlights the problems of doing science in the midst of a pandemic caused by a novel (new) virus: An enormous, unprecedented volume of studies have been published on COVID-19. But unfortunately, many don't hold up and are methodologically flawed. And, yes, the facts are not always in before some make some quick conclusions that are flawed.

"Under normal circumstances, years-long debate would filter wheat from the chaff, but everything is happening so immediately now," he said. Who is to know the truth if it cannot be proven.

No time for science to self-correct

David Kriebel, ScD, a professor of epidemiology at the University of Massachusetts-Lowell, has followed the controversy. While he agrees that the PNAS study is flawed, he does not agree with a retraction at this time. Regular people would have a problem with this logic. It is either right or it is not.

The paper wasn't a failure of the peer review process, he said, but rather a failure of understanding the limits of science during a pandemic. Regular Joe's will have a problem getting that nuance.

"The kind of science we are talking about—and the public has become so remarkably informed about—is applied science being used to inform decision-making on a mass scale," Kriebel said. "That kind of science is really quite different in important ways from the work of geologists, chemists, or astronomers. It has an urgency; it has to be translated to millions of people, and quickly."

Kriebel said that usually science is self-correcting, given enough time. Other scientists with enough time, over time get it right. But currently there is not enough time for science to self-correct when it's being used to craft public health policy. He said that's a problem for policy makers over-relying on "capital S" science to justify decisions.

"It's actually not helpful for scientists to hide behind a curtain of certainty when they do not have certainty. There is a lot of uncertainty about masks. But that doesn't mean we shouldn't be wearing them," Kriebel said. Instead of clamoring for scientific studies to back up mandates on mask use, Kriebel argues for more transparency in public health messaging.

"I would say, 'Mask use is our best judgment right now, and we will tell you if we get more evidence," he said. Of course that leaves a lot of room for the conclusions on mask use to be wrong but they won't be proven-out until a lot more time passes. But, the scientists are not given the time they need. Perhaps they should not make definitive pronunciations if they are not certain. Why they do is something explained only by other scientists that have eventually gotten it right.

Both Kriebel and Haber agree that masks ***probably*** do offer a level of protection, but right now there is no way to tease out how much protection masks offer versus physical distancing of 6 feet or more, or hand washing.

"The world is much messier than we would like to admit," said Kreibel. "We scientists do our best and we must admit our uncertainty."

One might conclude that no conclusion can be made but still most scientists seem to want to err on the side of caution. Does caution hurt more than it helps. Ask the man who wears a mask every day and finds out a year later that he did not have to do so. See what I mean?

Chapter 4 Reuse COVID Masks???

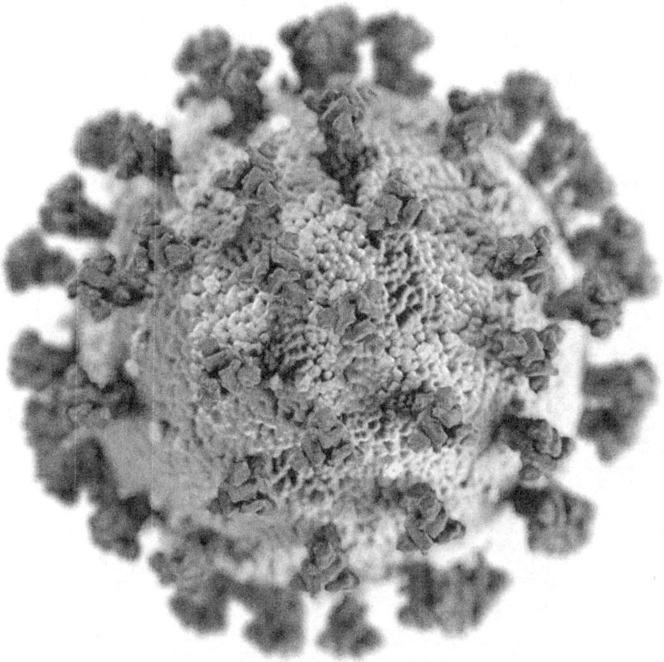

Is a mask still effective when your breathing makes it soggy?

I have a bunch of unsolved puzzles with my personal use of my face masks that I use to protect myself and others from contracting the virus. Let me ask you all a question.

Have you ever been in a doctor's or dentist's or other provider's office and engaged them in a conversation? Were they wearing soggy face masks?

Check out picture on next page

Depending on how late it is in the day and how frequently the provider changes their mask, have you ever noticed droplets escaping from the side of your doctor's mask, perhaps hitting you (unknowingly on his part) even as he spoke to you. If you have, you are not alone.

People who wear their masks for a long period of time will inevitably have a soggy mask. There is not much written about this phenomenon other than that it needs more scientific evaluation. Depending on the result of the investigation, it might render the use of a mask later in the day as being substantially less effective. That does not mean you should cut out the soggy part as the lady in the next picture has chosen to do. It may be more comfortable but it defeats the notion of the mask, kinda, don't you think?

All the science is not in folks no matter what CDC and their disciples tell you. Is it better for risk prevention to wear a mask rather than not wear a mask?

For full disclosure I am not a doctor and my research is my own and not necessarily based on the latest science. It is based on my thinking and the power of observation. Think about it and ask a trusted source of yours any question that this discussion brings to your mind. This is all my opinion. But it has value especially if you have not thought about it and the science from the CDC is clearly inconclusive.

The science says to change your mask once a day and do not reuse it. If it is a cloth mask, wash it and dry it and you may use it again. Now, here is the question? Why should you not reuse a mask? Answer. Because it may be laden with contaminants including the coronavirus and if the virus did not get you in the first wearing, it may get you the second day?

The virus that infects you more than likely would not come from your own breathing. It may not come from you. In fact it is highly unlikely that you would get the virus from your own inhalants of just your breath.

However, it may be a virus that was exhaled onto your mask from another person and it is on the outside of your mask such that on an inhale, it is brought inside your mask—even if you have not inhaled it.

If ten minutes after you put your mask on, the contaminant has entered the mask, should you not change your mask then? Of course you should but there is a problem in such a cavalier answer.

How can you know when a microscopic organism on your mask presents a threat to you. Scientists assume that by the end of the day, your mask is contaminated and you should dispose of it or clean it for the next use.

However, even Dr. Fauci does not knows when any mask becomes contaminated. More than likely, if it is infected, it got infected in what we might call as infection window. That would be seconds, minutes or hours after the beginning of the day or the same periods before the end of the day or the end of the period in which you wear that particular mask? Nobody knows is the point.

Do you see the dilemma? So, how often should you change your mask? Well, to be safe, you should change it every second or every minute but this is impractical and so such a method has been ruled out by expert scientists. They say change your mask every day. That is a reasonable amount of time even if it does not stop the bug from claiming you. Do you see the problem? That's why masks cannot be a 100% preventative for COVID-19.

Only a new kind of automated mask that has a windshield washer mechanism built in that could sanitize the mask continually throughout the day with whatever solution was needed, could be 100%

effective. But, no such mask exists and such a mask would be difficult to build and if built, it would be difficult to wear. Moving parts make noise and the mask noise could also create a distraction to the work you must perform every day.

So, despite what Dr. Fauci and others may say. the fact is that no mask cannot be 100% effective. Moreover, depending on the material of which a mask is composed, it has a definite impact on the effectiveness of the mask. How could Dr. Fauci know what material is in your mask? Consequently I am aware of no effectiveness charts on masks per se. Some materials are better than others but no mask solution can be 100%.

When you read about masks, you will find the literature suggesting that it would help to have a training course on the proper way to wear and handle a mask. I am not kidding.

Think about this in light of the fact that in the US, there are over 330,000,000 people of all ages. Think about training them all to wear a mask properly? That's why you see some people wearing bandanas and others paper masks and still others have the mask below their nose and still others take theirs off and put them on again throughout the day. This is not recommended but you can bet that most are unaware who have not gone through the training regimen.

Would the person who just read this last paragraph, who has seen the training guide or who has taken the course, please raise your hand.

Don't worry, I am not looking. But, you see the problem. Are masks 90% effective in the morning and 10% in the afternoon. I don't know. Looking for such an answer when the science is not in, all I can find in the scribes is that such answers are yet to be available. Masks thus at this time at best are imperfect instruments. Nobody has a number for how effective they are. Why the CDC and other organizations sweer by them is a conundrum. If the CDC offered proof, the puxxle would be solved.

There is no doubt that the best solution to prevent the virus is isolation. OK! I agree that is impractical. The next best solution is almost isolation, etc. etc. etc. Social distancing of 6 feet may work unless Charlie the Sneezer is in the next church pew.

Check out therapeutics like Hydroxychloroquine and Remdesivir and others. Used as a prophylactic, they may be lots more effective than masks.

Does the Sun purify a mask?

Folks, here is what I do. Don't do this yourself and consider it a scientific solution. It does work for me, I think. It is what I do and I have yet to get the virus but I may have distanced myself and may never have gotten a germ droplet on my mask. Anyway, I always thought this was a good method for me. So far, it has worked.

I have a cloth mask made by my sister, an N95 mask bought for me by my beautiful daughter, and I have a paper surgical mask bought for me by Katie, my lovely daughter,

I am never without all three of these masks when I drive anyplace. I keep them all inside out on the left side of my dashboard in my 1998 Ford Expedition. In one year, it will be a genuine antique. The most comfortable mask is the blue surgical mask. When I go to the store, I take it off the dash and put it on for my short trip – typically a half hour.

When I come back I put the packages in the back and climb in the drivers' side and put the mask inside out on the dashboard so that it does not blow away. On a sunny day in the summer, the mask gets the sun, though filtered by the windshield. It is often a day or more before I use the mask again under the same circumstances.

I thought that was all I had to do. But, there is always a gotcha. If the UV of the sun does not kill the virus, then it gets so hot in the car, that I figure the heat will kill the COVID-19. Like masks per se, I am not 100% sure either method gets all the virus on the mask if there is any. One thing is for sure. After a day in the sun and the heat, the mask is certainly not soggy.

Scientists would say my method should not work. Why not?

Here's the gotcha. Natural sunlight, which lacks UV-C, can't kill COVID-19. It is a fact. UV-C can get the job done but it is not in

Sunlight: The sunlight does contains UV-A and UV-B and these may make a small dent on a potentially infected face mask, but the scientists say that "it takes hours and hours" and doesn't completely sanitize like UV-C does. If you buy a UV-C light that might and should work. The heat should work also

Marshal Shepard writing for Forbes Science thinks sunlight and heat in a car can help a lot. Hot cars are one of the most dangerous places to leave a human being, cellphone or pet. The question is could that same car heat that makes a car so dangerous be useful in the fight against coronavirus? Here's an explanation, but before that discussion, this disclaimer is necessary: NEVER sit in a hot car or leave any person or pet in one.

A recent study published in the New England Journal of Medicine confirmed that coronavirus can linger on different types of surfaces for significant periods of time. For example, the study found the following "linger times:"

- cardboard shipping boxes - 24 hours
- plastics - up to three days
- stainless steel - up to three days.

If you are like most families, you probably have been ordering many things from online distributors that arrive in boxes or plastic containers. A good mitigation strategy is to spray them with disinfectant or leave them sitting around for some period of time.

However, others say that many of the media stories about virus lingering times neglected to mention that the studies took place at room temperature (roughly 72 degrees F).

It looks like the Coronavirus lives a few days on many surfaces at room temperature, but it dies much faster at higher temperatures. For example, if you heat a material up to 130 degrees F, which is the high setting on a dryer or a very low oven setting, you only need about 20 minutes to kill greater than 99.99% of the virus." At 150 degrees F or higher, it only takes about five minutes to kill the same percentage of virus. However, it may not be practical to throw your mask or package in the oven. The clothes dryer is a good alternative for the masks, but what about those packages? Spray them with disinfectant.

If you have a 1998 Expedition or another "hot" car available like I do. here is where your car and something called "vehicle heating dynamics" can come into play. The fact is that it is brutally hot inside a car left in the sun. Dark materials (dashboards, seats, and other objects) heat up significantly and then heat the air through the processes of convection and conduction. The objects also emit infrared (heat) energy. This automotive "greenhouse" effect could be useful in killing coronavirus according to scientists.

There are a lot of tests that show Coronavirus does not withstand heat very well. The infection on a mask can be completely killed or 99.9% killed. It is important that only inanimate objects such as your mask be considered for this practice. Dr. Calvin Mackie, a renowned engineer and STEM ambassador--the founder of STEM told colleagues "I keep my mask on the dashboard of my car for this reason." Hey did I not say that a while back? That's what I do and I need no brown bags.

What about a sun-baked car?

The fact is a sun-baked car can help and would help and I am glad because that is what I do. Some believe that leaving a face mask on a car's dashboard may magnify the sun's ability to disinfect. But again, UV-A and UV-B rays won't do much in this regard and Anderson said the windshield would actually reduce exposure. But, if you get the heat, you've got the virus killer.

So, what could potentially kill the virus is how hot the car gets. SARS coronaviruses aren't heat resistant. but this method isn't scientifically proven and he worries this tactic would give false hope. Think it through fols. It does make sense.

Despite the fact that it has worked, scientists suggest that when you're done wearing a surgical face mask or respirator, throw it away and wash your hands. Never reuse it.

Wash your cloth face mask after each use.

The average masks you can buy from a local drugstore aren't seen by experts as being enough to filter out viruses.

For that purpose, experts recommend special masks with fine mesh that can capture very small organisms. These also have to be worn correctly for them to work.

Masks worn over the face are also unable to protect you from getting airborne virus particles, from a cough or sneeze, into your eyes.

When it comes to the flu, prevention is still the best method of keeping yourself safe from contracting this highly contagious virus.

Chapter 5 Counterfeit Masks 'R Us

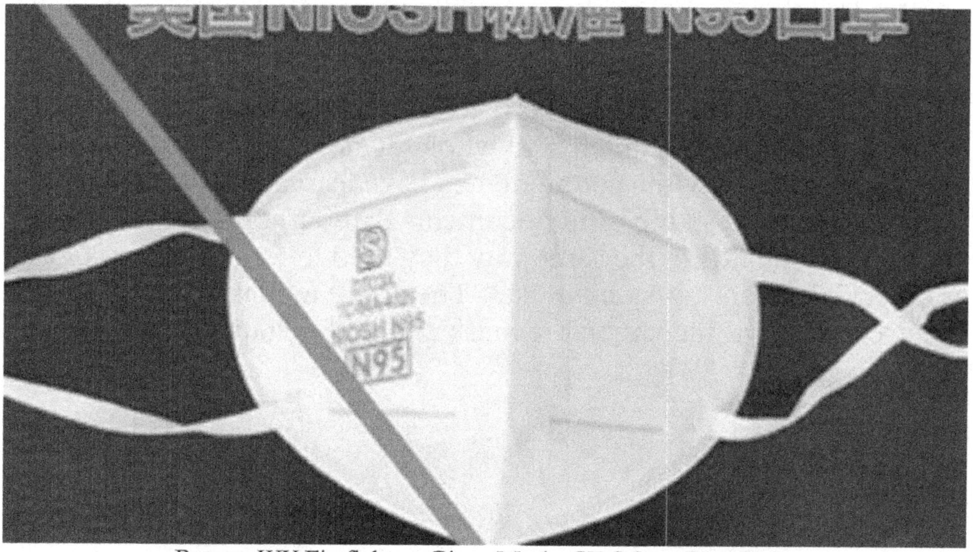

Report: WV Firefighters Given Masks CDC Says Were Fake

The Centers for Disease Control issued a warning that a model of masks was counterfeit, but West Virginia officials still distributed them to state first responders. Supposedly, like a rag placed loosely on your face, they are better than nothing, but not as good as they are supposed to be.

Masks are the latest craze for small and even larger sewing shops around the world. All you have to do is talk about a shortage of a product that is easy to make and the world's artisans will be baking it or making it. The word on the street is the mask counterfeiters have total disregard for people's lives. Quality is not in their dictionary so be careful. The new mask manufacturers have flooded the world with hundreds of thousands of fake masks while the needs of those suffering from coronavirus concerns have depleted the world's supplies.

No, I am not kidding. Counterfeit masks are being shipped worldwide. They look good but they are not necessarily of the highest quality material or workmanship. Worse than that, they are made across the world often in unsterile sweatshops. They are often creative mimics of famous brands with names of well-known medical supply companies and manyt come with forged certification stamps.

To repeat, producers of the counterfeit masks are labelling their wares with fake brand names of well-known medical supply companies such as 3M. Hundreds of thousands of counterfeit medical masks are being peddled as the world grapples with the coronavirus crisis.

Certification is important for masks to be able to sell to meet the demand of legitimate retailers and wholesalers—especially to the medical industry. So, the counterfeiters have learned to even forge "CE" certification stamps and documents that purport to attest to European standards in the same way they used to forge Ralph Lauren or Gucci labels on shirts and purses. The desire to make a buck supersedes everything else in the mask counterfeit industry. Business first!

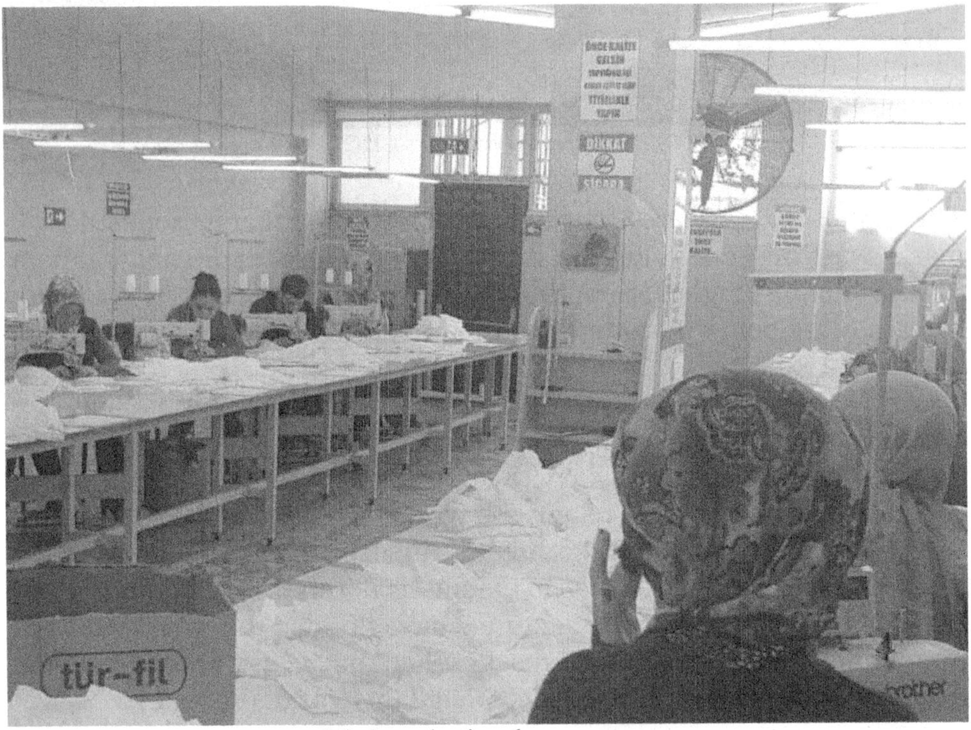

Mask production shop overseas

There are a many photos in various magazine and Internet articles that show workers in various textile factories across the world such as Turkey, one of the major sources of counterfeit products. Many shops stitch together medical masks in conditions that experts would describe as unsterile and substandard.

There is one report that just recently Turkish police seized 1 million masks and they arrested five people in a raid on a sweatshop making unauthorized medical supplies in Istanbul. Unfortunately, the problem is a lot bigger than that.

The companies use whatever tools they need to make their wares appear to be the real deal and as legitimate as the highest certified masks in the industry. For example, they have no problem using Photoshop or whatever cheaper tool they can find to make their own certificates.

Industry insiders stumble across opportunities to purchase these masks and they have a tough time differentiating the sources. The price often convinces merchandisers to buy the inferior brand for distribution.

The counterfeit protective mask articles often wind up being purchased for industries such as the electronics industry, not just the medical industry. They are even enlisting box manufacturers to make fake boxes with logos on them just like they were selling fake Nike T-shirts. Masks fill in the cut and sew market quite well since nobody's buying T-shirts any more. The same companies that made knock-off T-shirts have naturally begun to manufacture masks."

Authorities try to do something but there are a ton of sources out there. There is a report that just last week Interpol warned of a spike in fake masks and other medical supplies, describing a 90-nation operation that resulted in 121 arrests and the seizure of tens of million of dollars worth of "potentially dangerous pharmaceuticals". Masks are just another product in this illicit marketplace.

Even without wearing masks to conduct their activity, criminals stop at nothing to make a profit." It is their livelihood and they are in it in a big way according to Interpol. "The illicit trade in such counterfeit medical items during a public health crisis shows their total disregard for people's wellbeing, or their lives."

With so many countries and territories and states within those countries making masks mandatory, the demand is unprecedented. With people using more than one mask a day and throwing it away, the demand increases by several X factors. And, so, surgical masks have become a crucial component in the worldwide battle against coronavirus. The industry is so big now in such a short time, that politicians will be hard pressed to ever suggest that masks can go back to being optional—even when the pandemic is over.

The mask has in fact become the most visible icon of a pandemic that has dramatically transformed life across the planet. Life is not the same as it was in February 2020 and if the illegal mask purveyors have their way, it will never be the same. Medical workers across the world have complained they don't have enough and it is difficult with illegal certifications to tell the wheat from the chaff in terms of masks.

Physicians and nurses from Mozambique to Morocco and from Iran to Italy to the United States claim that they are forced to reuse masks that are ordinarily thrown out after seeing each patient. National and local governments are scrambling to obtain more masks. The new backdoor industry fills the bill to an almost unsuspecting public. Now you know!

It makes it tough for medical professionals and policymakers who would have it tough finding legitimate products on a good day to face the added complication of potentially dangerous fake masks adding to their list of challenges. Legitimate medical product manufacturers and brokers worry about their reputations and the spread of the COVID-19 increases the more disease is spread via the use of defective masks made with shoddy materials.

It is a double whammy as the demand increases, it is being met by more and more illegitimate products. Protective masks, which can be used in medical as well as industrial fields, have become a coveted commodity for frontline hospital workers seeking to protect themselves from contracting Covid-19. At the same time, ordinary people are simply seeking to stem its transmission.

The fake masks may pose a threat to everybody including the wellbeing of medical workers. Why? It is because they are not made with the correct materials or in sterile environments. What a shame to get the coronavirus from the very mask one uses to seek protection. It would be OK otherwise as these are not pharmaceuticals, they are masks. The problem comes about because they lack the specifications that would prevent viruses or bacteria from entering into breathing passages, according to industry experts.

This means that the end product is a mask that does not fulfill the density requirements of the certification. This of course means the virus can come through to the nose or mouth of the person and infect them or release the droplets that infect somebody else.

Medical industry executives understandably are very upset as this may bring eventual safeguards that will put some legitimate companies out of business. There are some who don't get it and won't get it however as they persist in making inferior products. For example it is commonplace to hear about some fake masks produced by textile firms rather than healthcare manufacturers. They are deemed "better than nothing".

Counterfeiters say that as long as they comply with basic standards of use—three layers of material and weigh about 60g, they'd likely be enough to prevent the infiltration of coronavirus. The virus of course travels through droplets of liquid. At the very least, it can be said that more academic studies should be conducted. Some in the industry think that the whole legitimacy issue is because the big players want all the business themselves. It's tough to know.

The reason demand appears to be met in this period of tremendous global demand is because of new entrants, many of whom are not on the up and up. However, the fact is that there is no way traditional medical manufacturers could produce enough of the masks. Industry spokesmen have noted: "It's an urgent situation. Not a normal time."

The extent of the rogue production is unknown but it is known to be widespread. Industry sources say it encompasses factories in China, India and Turkey that have been panicked over the collapse of the textile industry in the wake of the global coronavirus pandemic. Just one factory in Turkey, it is said produces about 450,000 masks a day, and there are likely dozens of such places in the country.

A recent report cites a fact that China's government contributed to the mask chaos by ordering companies that knew little about medical standards to produce and procure as many masks as possible. They reached out to brokers across the world who put in orders for tens of millions of masks. This created a "shortage" and it prompted lower quality masks to be produced globally.

There is a cited statistic that Turkey, a major mask producer ordinarily would produce about 150 million masks a year. A re[ort tells us that China apparently ordered 200 million masks from a Turkish manufacturers in January, with orders also coming in from Italy, Netherlands, France Poland and Germany.

This created an unprecedented surge in demand that under regular circumstances could not be met. Additionally, the shortage drives up prices. Prices in Turkey spiked from about $.02 to $.04 per mask to $.50 to $.60. Overall the price of masks keeps going up and the counterfeiters get their share from the lower tier price points.

One things is for sure. Counterfeit masks will be here for quite awhile.

Chapter 6 Care for Your Mask

You probably have heard or seen in print the wise crack advice for those asking about how to take their mask off in public: "Don't ever take your mask off in public." But, we know that is impossible so we offer a few suggestions how to do it more safely.

For example, I may ask: "If I need to take my mask off in public, where do I put it until I can put it back on? What if I need to eat or drink something while wearing a mask?

Unfortunately, there is limited research in this area as with many other questions about masks. The CDC does not do research on all aspects of mask usage. However, the scholarly advice is to think of your mask as a part of your face: That means you should wash your hands before touching the mask and gently remove it only by the ear straps. That means you should keep your "paws" off your mask.

Before you leave the house, when you pick up your mask, bring something to store it in case you take it off for awhile. For storing it, the recommendation is to have a designated brown paper bag with you that you can place it in when not using it. Don't just set it down. If you are in public daily and will be wearing the mask for hours, change your brown paper bag daily. Quite frankly, it is a good idea to dispose of your wet soggy mask after it reaches that point and have another clean one available when you have to mask up again.

Before you take the bag out of the house, label one side "outside" for the outer side of the mask (the side that faces the public), and label the other "inside" for the inner part of the mask (the side near your mouth). Always put the mask in the bag with the inside part corresponding to the side marked "inside" so that you don't contaminate it with what is on the outer-facing part of the mask.

Here is the question I answered before but it is worth another go. "My mask is getting wet from condensation from my breath. Do I need to change it out?"

In healthcare, using disposable masks helps with this problem, but if you are using a cloth mask and find that this happens often, determine if the cloth you have is breathable or too thick.

The CDC suggests using a T-shirt, likely cotton, for DIY masks. However, I have an NBC story that talks about a study finding other fabrics that may work better—this is an example of an area where more research is essential but the fact is research on masks is still quite scant.

Sweat absorbable or moisture-wicking fabrics similar to Dryfit have not been tested, but they may help absorb moisture from breathing. Double check before you deploy anything untested.

Avoid waterproof materials. It stands to reason that if fluid can't flow through them then you probably won't be able to breathe through them. If you could see the sides of such a mask in use it would be like a wind tunnel. The inside would get soggy quite soon and overall the mask would be ineffective.

The truth often not disclosed about wet masks is that there has been studies. One study that looked at the effectiveness of cloth masks notes the moisture inside of the masks as an issue and it provides major risk for germ transmission. It stands to reason.

I keep wondering about the fact that it is recommended to wear a mask no longer than one day and to not put it on the next day without washing and complete drying. If it is a sure bet that the mask is contaminated after one day, it is also a fact that it did not reach contamination level the instant that you took it off for disposal or washing.

And, so, because the experts have no idea how often you should really change a mask in a given day, they avoid the question and say take it off and don't reuse it at the end of the day. Surely if it gets contaminated it is sometime between the first hour and the last hour of use. It is just an unknown because nobody has done the hard research. So, for sure if your mask gets soggy, get it off for a new one and check out the material as it may not be a legitimate mask.. It may be counterfeit and made of poor materials.

Another question is "How often should I wash my mask? Do I need more than one? Surely you would have these questions if you gave your mask more thought.

These questions deserve further scientific investigation and direction by the CDC, but until we have that guidance, I would suggest taking a logical approach. These are not recommendations because I am not a doctor but they are what I do so take it for what it is worth. The CDC offers no guidance here.

Ideally, at a minimum, masks should be changed after every wear, but this may not be possible for most people. Assuming no one in your home is sick, it depends on how many times you leave your home each day and each week how often you change your mask. Putting on a contaminated mask has some risks but like all things having to do with masks, not putting it on is deemed by the experts to be riskier.

Try to have at least two masks or even more for the times when at least the prior use mask is being laundered. How often you change it also depends on how long you're wearing it and where you go in public. If you are among people in a crowd for example and they are not social distancing, the fact is that your mask will more than likely get contaminated sooner.

Everyone should be practicing social distancing as in my humble opinion (not necessarily an expert opinion), other than prophylaxis therapeutics such as hydroxychloroquine, the closer social distancing gets you to an isolation distance, the safer you are. I am not being cute. That is a fact.

Be smart and recognize that in places like grocery stores, you might have to engage with people in aisles in closer proximity. So, you may want to change your mask sooner if you have a change available. Have as many masks as you can afford and you can manage. It won't help for you to have so many masks, that you can't tell which is next in rotation for use and for cleaning.

To avoid potential contamination from the mask, I suggest washing and drying it as soon as you get home, especially if you don't know the next time you'll need it. This prevents you from having a potentially contaminated mask lingering around your home.

I keep three inside-out masks on the dashboard of my car and I replace them as I think is appropriate. In hot weather, I think they are as purified as putting a mask in a clothes dryer. This eliminates the brown bag for the most part for me as my trips are short and I am not away for an extended period of time—where taking off the mask and putting it on is a requirement.

Does it matter what detergents I use to wash my mask? One would expect the CDC to offer guidance on this but they do not.

This question does deserve CDC research and guidance, but until then, we can only rely on limited information from studies. For example in studies about irritation from contact dermatitis, there are some findings. Many detergents can cause rashes, and most detergents have fragrance and enzyme ingredients that can cause pore-clogging or breakouts, so the best option is to use detergents free of fragrance and enzymes. Do research to make sure you have the right soap or detergents.

If you launder your masks with the rest of your laundry, rinse it for extra time in plain water to ensure fragrances are removed. You can also spray it with a 5% bleach solution (most bleach is 5%) and let it dry, but test this first to be sure it doesn't irritate your skin. Again as I said earlier in this chapter, treat your mask as if it is your face with the same sensitivity as your skin.

Be careful about chemicals, such as fabric softener or bleach. Make sure you eliminate this possibility. Don't use extraneous chemicals in sterilizing or washing your mask?

More than likely, whatever detergent you use for your laundry will be OK for your mask. But, it would help one time at least for you to check it out so you are not surprised by getting an irritation on your face. Remember drying a mask in a dryer helps remove infection also.

As noted previously, treat your mask as if it has the same sensitivity as your skin since it will be up against your skin for long periods. It's best to avoid fabric softeners, bleach, and any ingredients that might cause an allergic reaction, rashes or other symptoms, including fragrances that can cause headaches with long exposure. In addition, do not spray down masks with products such as Lysol and other disinfectants since their labels state that they may irritate the skin.

Here is another good question. "Should I dry my mask in the dryer or air-dry it?" The answer is yes either way is OK, but do not add other chemicals such as dryer sheets. Air drying is fine but if the wash did not remove infection, air drying won't either.

How about this question? What does it mean to "fit snugly but comfortably against the side of the face" against your face?

Check how your mask looks. You don't not want to see redness or deep lines in your face as a result of wearing your mask. If you are a bruiser, take that into consideration. Some people bruise more easily than others, but if you see any imprint in your skin, you should loosen your mask.

There should be a full "cupping" around the nose and chin area with minimal space or opening on the side. If the mask makes it hard to breathe because it is pushing against your nose, loosen your mask. Be careful because skin breakdown and pressure sores can occur if the straps or mask is too tight. Furthermore materials such as rubber bands create tension and may create unforeseen problems.

Mask effectiveness: Will my mask catch all the droplets when I cough and sneeze? The CDC can answer and has answered this. The answer is a big fat "No." Don't expect it. The answer is to still cough and sneeze into your elbow.

If you could wear an iron mask pressed against your face, it is certain nothing would get through as it would be 100% effective. Unfortunately, after about five minutes or so, you would probably die because you could not breathe with an air-tight iron mask. Mask fabric must be somewhat porous for you to be able breathe. Consequently, droplets can and will still travel through. It is by design so no mask can be 100% effective. The mask helps decrease the amount of droplets that would travel from your mouth or nose if your face were uncovered but it does not stop them all.

The mask may also help reduce droplets that would travel from others to your face, though the existing evidence for preventing infection this way is weak unless someone is directly coughing into your face.

You can see from this article that masks are not 100% effective and they cannot ever be. Moreover, you cannot find the effectiveness of a mask that is worn by you because as the face on which the mask is affixed changes, the effectiveness change. Consider that when you think you are safe. That's why a lot of people have declared masks as pure bunkum yet the experts suggest they are still better than nothing.

Who should not wear a mask? The CDC currently recommends not using masks for children under 2 years old. Don't use a mask for anyone who has trouble breathing, or is unconscious, incapacitated or otherwise unable to remove the cover without assistance.

Here is another question: If you have a pre-existing condition, such as asthma, that already makes it difficult to breathe, will wearing a mask make it worse?

More than likely yes but there is not enough evidence to know the effects of wearing a mask in people with asthma, but having asthma is often associated with a higher risk for anxiety and for allergies, with one condition sometimes triggering the other.

It is possible, but not shown in research, that wearing a mask could affect the relationships between those conditions. In other words, you are on your own so make sure someone in the family knows about your concerns if you have asthma, before you mask up.

Until we have more data regarding masks and conditions like asthma, people should follow their best judgment. If your anxiety, asthma, allergies, or a similar condition makes it too difficult to breathe while wearing a mask, remove it immediately and speak to a healthcare provider.

What if I need to wear oxygen. Do I put my mask over the oxygen tubes? This is another question without an answer from the CDC.

It requires more research and guidance from the CDC. If you choose to wear a mask, while receiving oxygen through nasal cannula, be aware that the SARS-CoV-2 virus can remain infectious on plastic, so the outer part of those tubes should be cleaned every time someone returns from having left their home. It is another source to pick up the virus.

What should I do if I develop a rash from wearing my mask?

Assess the seriousness. Ask for help from your doctor or go to urgent care and get a professional opinion. Other thoughts are to wash your hands, remove the mask at home by the straps, wipe your face with a clean rag and allow your skin some time to recover.

Facial moisture is also the cause for certain skin breakouts. While not yet studied in everyday people, personal protective equipment (PPE) (masks) have been shown to cause contact dermatitis in up to 33% of healthcare workers. That's a high percentage so expect some irritation.

The risk of getting a rash depends on the multiple factors, including how long you wear it, the moisture produced, and your skin sensitivity to certain materials.

Will I get more pimples from wearing a mask? Sorry but like many questions there is no definitive answer to this. We don't yet have enough evidence to answer this question for everyday people. If your skin is prone to breakouts because of dryness or moisture, evaluate your face before and after wearing your mask and take note of changes in your skin so you can adjust mask use as needed. Remember that the more you stay home, the less often you need to wear a mask.

How do I protect myself if someone sneezes or coughs on the outside of my mask? It is already too late. But, you should immediately turn away from the person and remove the mask by the straps. If you are in public, discard the mask immediately to avoid transmitting potential germs on the outside of the mask to yourself or others. Throw it away if it's disposable or you can make another. If it's a cloth mask you need to keep, place it in your paper bag. Throw the paper bag away after this use. Wash your hands immediately after removal. Remember to continue social distancing while wearing your mask.

If I am not Santa Claus, must I shave my beard to safely wear a mask?

This question has no CDC answer. The Occupational Safety Health Administration does not require shaving beards but does state that a beard should not interfere with obtaining a complete seal with the mask. If your beard is exposed from outside the mask, there is not a good seal for the nose and chin areas, which should be covered. The answer then is to shave.

How do I handle my glasses persistently fogging up? Glasses are an issue and can be a concern. If glasses keep fogging up, it could mean the mask is not secure enough snugly around the nose area. Unfortunately, as someone who once wore glasses, the fog is sometimes unavoidable while breathing, even with a good seal. It's similar to wearing a scarf in the winter.

If the fogging of glasses interferes with sight, wash your hands and remove the glasses to clean them. I suggest using disposable eye lens wipes or, if using a reusable anti-scratch cloth, be sure to wash it daily if you need to clean your glasses throughout the day. Do not set the wipes down on surfaces you haven't cleaned and dried first.

I sure hope this helped. The information in this chapter came from various sources as well as personal observation and opinion.

Chapter 7 Why People Get Upset Over Wearing Masks

Is the mask the new symbol for tyranny in government

On May 6, 2020 Seth J. Gillihan Ph.D. posted an article intended to help us all understand the masking controversy. He titles his article *Think, Act, Be Why Are Masks Triggering Conflict and Rage?*

Dr. Gillihan leads off by suggesting that face masks have become a contentious issue in the debate around Covid-19. He's very right. His article forms some of the basis for this chapter. Our thanks for his insightful work

Since March, the government has been playing rope-a-dope with the freedom-loving American people—those more concerned about constitutional freedoms disappearing from tyrant decrees than catching the COVID-19 bug.

The people unanimously signed early on to give up all freedoms for a two week lockdown to flatten the curve so hospitals could be assured the resources to handle the anticipated rise of virus cases. Government took a lot more authority than given and that is why we have the controversy.

Nobody of whom I am aware, decided to give government carte blanche forever lockdown / shutdown power. Nobody of whom I am aware gave up on the Constitution so some governor could act out his or her secret life as a tyrant.

With the media having joined the Democrat / leftist camp and vice versa, there are not many strong voices complaining about the takedown of freedom to appease the COVID-19 lockdown crowd. Yet, many people are very upset.

The mask has become a symbol of the tyranny that is being foisted on Americans by the Andrew Cuomo's, the Phil Murphy's, the Tom Wolfe's and other leftist governors who want the country to fail. In my state, Pennsylvania the rope a dope, on again off a gain shutdown has caused a major backlash. Regular people such as me are irate.

Articles of impeachment against the governor have already been introduced by the legislature. In a recent letter to Governor Wolf, Rep. representative Daryl Metcalfe, the impeachment author, urged the governor to resign or face impeachment and possible removal from office. There are lots of reasons including the deaths of many senior citizens who were not adequately protected by the governor in nursing homes.

Many Pennsylvanians feel like we are living in Hitler's Germany or Stalin's Russia. Wolfe made his lockdown more severe in July with a broad announcement that included mandatory mask wearing and restaurants permitted to use just 25% of available seating and no counter seating like bars.

The reduced capacities at eating establishments is especially burdensome and is putting businesses out of business. Restaurant business owners reacted: "Reduction to 25% capacity is essentially the same as eliminating indoor dining entirely, and there is no scenario that restaurants can survive at that level of occupancy," said John Longstreet, president and CEO of the Pennsylvania Restaurant & Lodging Association, which is instead advocating for shutting down problem establishments. Wolf, an autocrat who belongs in seventeenth century England takes advice from nobody. It's his way or the highway.

Some things seem eternal when they seem to last forever, On Sunday, April 5, 2020, four months ago, Dr. Annie Bukacek, smelling a dead fish and still wearing a white doctor's coat and a pink stethoscope draped around her neck, stood behind a makeshift pulpit against the plain backdrop of a Hilton Garden Inn conference room in Kalispell, Montana.

She was there to deliver a sermon of sorts to members of the Liberty Fellowship, an anti-government, anti-globalism church led by a pastor named Chuck Baldwin that serves as a beacon for the extreme political right — a mix of constitutionalists, militia members, and separatists — in Montana's Flathead Valley. Nothing is needed more than somebody to speak out against the unholy consortium of the press and the progressives to shut down the country to hurt the president's election chances. They don't care about America.

Based on responses from a Facebook post and her own experience as a doctor, Bukacek made her case to the congregation: that the "alleged death rate" due to COVID-19 has been significantly inflated in order to justify otherwise unjust stay-at-home orders. Why would the government do this. Answer: *to hurt Trump.* The truth seems to be a major casualty of the pandemic.

Doctors, Bukacek argued, rely on "assumptions and educated guesses that go unquestioned," which has resulted in an overreporting of deaths. "Based on inaccurate, incomplete data," she said, "people are being terrorized by fearmongers into relinquishing cherished freedoms." (The most recent CDC data suggests that coronavirus deaths have actually been dramatically underreported.) All of the negatives including the negating of the benefits of hydroxychloroquine in fighting the bug are highlighted to hurt the president's activities.

Bukacek argues that because many who have succumbed to the disease also had other health concerns, they thus died with COVID-19, not of COVID-19. (People with "comorbidities" like hypertension, diabetes, or immune deficiencies are indeed more likely to die from COVID-19, but the infection itself is the trigger for the respiratory or organ failures that actually cause death.) The truth is not reported because it is perceived that it will help the president get reelected What a shame...

The Doctor is not a nobody. In fact she is a member of the American College of Physicians, Montana Chapter. She was voted the Best Family Physician in the area in both 2012 and 2019. And, most importantly, she is currently serving as a member of the Flathead County Board of Health. She thinks it is a ruse by leftists to take over that which they cannot win by elections.

For this growing number of skeptics, the experts, institutions, and authorities of mainstream science and medicine can no longer be trusted. And that underlying distrust makes someone like Bukacek — a doctor who positions herself as a rebel unafraid to speak (alleged) truth to the medical establishment — a potent symbol to people who are already nursing their own suspicions.

That's why Governor Tom Wolfe of PA is being impeached and why the people of PA can't wait to get him out of office.

Bukacek is not just speaking to a congregation in a conference room in the corner of Montana. Like the president, she's speaking to a frustrated, confused, and receptive audience of millions. A little truth would go along way today. Now, what is the real truth about face masks and is the loss of freedom worth the potential gain?

We might say that "the coronavirus pandemic is the Medical Conspiracy Super Bowl," the moment the left has been waiting for.

Controversy over the nation's governors' response to Covid-19 has intensified as states have begun tightening their once loosened social distancing restrictions. In this context, the wearing of face masks as discussed has become one of the most contentious issues. For one, the science cannot say that face masks are a 100% solution.

Nonetheless leftists have embraced mask-wearing as if it is a god-imposed requirement. They'll punish anybody who does not follow their creed. For example, there was outrage at Vice President Mike Pence for choosing not to wear a mask when he visited hospital staff and a patient at Mayo Clinic.

Anger has also filled online bulletin boards in the suburbs of Philadelphia, with people posting rage-filled comments about neighbors not wearing masks in public. Grocery store employees have been verbally assaulted for asking shoppers to wear masks. A man at a Dollar Tree store wiped his face on an employee's shirt when she asked him to wear a mask, and there has been at least one mask-related murder.

What is it about a seemingly neutral piece of protective gear that can be so inflammatory? How about because in the chapters in this book we have proven the CDC does not have all the answers and the effectiveness of the cure is anywhere from 10% up and with all the holes in the efficiency testing, the highest percentage may be less than 50% on a good day. But what angers the left is that they think it is a 100% solution and the right is trying to kill them by not wearing masks or something like that.

Masks are clearly tied to the political conflict over our response to the coronavirus and the attack on our freedoms from the left. Those who lean left politically tend to see the virus as a more dire threat; those on the right are more likely to downplay its seriousness or compare it to less deadly strains like the flu, often following the lead of conservative politicians. The right is more concerned about the Constitution and freedom whereas the left wants the left to win all ideology battles even if the solution violates the US Constitution.

Masks are also viewed as a visible marker of political loyalty, triggering feelings of *us-versus-them*. A politically liberal person may assume that someone wearing a mask is "on their team," while those who don't wear masks must be Fox News-watching Republicans. The anger they feel is not simply about the mask, but about believing the non-mask wearer is a certain type of person whom they have been trained by the media to dislike.

On the flip side, the politically conservative might interpret calls for masks as politically-driven efforts to play up the seriousness of the coronavirus. Being asked to don a mask then becomes not just a request to protect the health of others, but to give up their worldview and political allegiance and their freedom to make their own decisions in life.

The wearing of masks has clearly become a catalyst for political conflict, an arena where scientific evidence is often viewed through a partisan lens. Most Democrats support the wearing of masks, according to a poll conducted by researchers at the Pew Research Center.

Most Republicans do not buy it. Many have checked the analyses done and they know that masks are breathable by design and this makes them ineffective by design. Some say the Republicans are following the lead of the president: Trump has been reluctant to wear a mask, saying that it did not seem right to wear one while he was receiving heads of state at the White House. He put a mask on in public for the first time during a visit to a military hospital to satisfy the press. It may not help and it is certainly an annoyance.

The battle over masks has escalated during the final weeks of the campaign season. The general election is in November, and activists in both parties, Republican and Democrat, are working feverishly to ensure victory at the polls. Some of them have faced off on the issue of masks: as Timothy Akers, a public-health professor at Morgan State University, a historically black college in Baltimore, says: "We're seeing politics and science literally crashing."

The dispute over masks embodies the political dynamics of the campaign. It also reflects a classic American struggle between those who defend public safety and those who believe just as deeply in personal liberty and constitutional freedom.

Masks are not the only thing that have become controversial. The left says a drug used by hundreds of millions of people for 65 years to fight malaria will kill you if you take it, while doctors say it saves their patients lives. Why is this? One reason, President Trump said it works. The next few chapters will address some of these other phenomena that the left dismisses. We've unmasked enough information on masks in the prior chapters to give us all a lot of thinking material.

God bless you all and Thank you.

Chapter 8 The Hydroxychloroquine Saga

Tell me about the problem?

Javits Center, NYC conversion to a 2900 bed hospital

An international poll of more than 6,000 doctors released several months ago a study that found that the antimalarial drug hydroxychloroquine was the most highly rated treatment for the novel coronavirus. You'd never know what if you read or listen to the lying media.

The survey conducted by Sermo, a global health care polling company, of 6,227 physicians in 30 countries found that 37% of those treating COVID-19 patients rated hydroxychloroquine as the "most effective therapy" from a list of 15 options.

Of the physicians surveyed, 3,308 said they had either ordered a COVID-19 test or been involved in caring for a coronavirus patient, and 2,171 of those responded to the question asking which medications were most effective.

Sen. Ron Johnson, Wisconsin Republican, sent President Trump a letter Friday from more than 700 physicians urging him to expand the use of hydroxychloroquine for coronavirus outpatients by removing federal and state restrictions limiting the drug's use to hospitals.

He said he forwarded the request to the White House after gaining support from 776 physicians, just 14 hours after he began circulating the letter, which asks Mr. Trump to issue presidential directives allowing doctors to "fight with all the weapons we have at hand."

The US Center for Disease Control has been working overtime for most of 2020 and part of 2019 in response to a pandemic of respiratory disease spreading from person-to-person. It has been caused by a novel (new) coronavirus. There have been other coronaviruses but this one is different and it has received more attention than any virus in recent history.

The disease has been named "coronavirus disease 2019" (abbreviated "COVID-19"). It comes with a very serious public health risk. The federal government of the US is very concerned for the safety of US citizens and has been working closely with state, local, tribal, and territorial partners, as well as public health partners, to respond to this bleak situation. COVID-19 can cause mild to severe illness; most severe illness occurs in adults 65 years and older. For some reason it creates a fatal situation in many adults over 65 years of age.

The United States is the leader of cases in the world

It began slowly in the US and as soon as President Trump saw its lethality. And its origin in China, he cut off all travel to China and later Europe and then Great Britain in an attempt to keep the virus from our shores. Different parts of the US have been experiencing different levels of COVID-19 activity.

The United States nationally is coming out of the acceleration phase of the pandemic. The duration and severity of each pandemic phase can vary depending on the characteristics of the virus and the public health response. It seems like the model projections have the infection and death curves flattening meaning the virus has been substantially but not entirely mitigated. CDC and state and local public health laboratories have been testing for the virus that causes COVID-19.

CDC Recommendation

With over 25,000 deaths already in the US, and the cases increases substantially in recent weeks, The CDC suggests that everyone can do their part to help the country respond to this emerging public health threat. Here are some of the CDC's important guidelines to which all Americans are asked to adhere.

Part of the CDC recommendation is for individuals to use a cloth face covering to keep people who are infected but do not have symptoms from spreading COVID-19 to others.

The cloth face cover is meant to protect other people in case you are infected and it offers some protection for you from others. The cloth face coverings recommended for the general population are not surgical masks or N-95 respirators. Medical face masks such as N-95 are critical supplies that should be reserved for healthcare workers and other first responders, as recommended by CDC. If the supply of the N-95 were adequate in the US, there would be no restrictions on the type of face covering for individuals who are not part of a medical team.

The cloth face cover is not a substitute for social distancing, the primary means that the US is deploying to stop the spread of the virus. The CDC continues to recommend that people try keep about 6 feet between themselves and others.

The White House's "Slow the Spread" guidelines came after the 15-day shutdown and are in place until April 30. These are part of the nations effort to slow the spread of COVID-19 through the implementation of social distancing at all levels of society.

People 65 years and older and people with severe underlying medical conditions should take special precautions because they are at higher risk of developing serious COVID-19 illness. The death rate is substantially higher for elderly patients.

If you are a healthcare provider, use your judgment to determine if a patient has signs and symptoms compatible with COVID-19 and whether the patient should be tested. If there were enough tests for everybody more people would be advised to get tested but the tests are in short supply but this situation is getting better.

Those people who get a fever or cough should consider whether they might have COVID-19, depending on where they live, their travel history or other exposures. All of the U.S. has been seeing some level of community spread of COVID-19. Testing for COVID-19 may be accessed through medical providers or public health departments, but there is no treatment for this virus. Hydroxychloroquine has shown some promise as a prophylaxis and as a cure for the virus but there are no confirmed studies of this as of yet.

The good news is that most people have mild illness with the disease and are able to recover at home without medical care. Some have major illnesses however, and need hospitalization.

If you are returning from a country with travel restrictions, you should stay home and monitor your health. All other international travelers please follow CDC instructions during this time. Your cooperation is integral to the ongoing public health response to try to slow spread of this virus.

The complete clinical picture with regard to COVID-19 is not fully known. Reported illnesses have ranged from very mild (including some with no reported symptoms) to severe, including illness resulting in death. While information so far suggests that majority of COVID-19 illnesses are mild. About 16% of the cases reported from China were very serious.

A CDC Morbidity & Mortality Weekly Report that looked at severity of disease among COVID-19 patients in the United States by age group found that 80% of deaths were among adults 65 years and older with the highest percentage of severe outcomes occurring in people 85 years and older. People with serious underlying medical conditions — like serious heart conditions, chronic lung disease, and diabetes, for example — also seem to be at higher risk of developing serious COVID-19 illness.

The COVID-19 is a pandemic. FYI, a pandemic is a global outbreak of disease. Pandemics happen when a new virus emerges to infect people and can spread between people sustainably. Because there is little to no pre-existing immunity against the new virus, it spreads worldwide.

The virus that causes COVID-19 is infecting people and spreading very easily from person-to-person. On March 11, the COVID-19 outbreak was labeled a pandemic by the WHO.

This is the first pandemic known to be caused by this new coronavirus. In the past century, there have been four pandemics caused by the emergence of new influenza viruses. As a result, most research and guidance around pandemics is specific to influenza, but the same premises can be applied to the current COVID-19 pandemic.

Pandemics of respiratory disease follow a certain progression outlined by the CDC. They begin with an investigation phase, followed by recognition, initiation, and acceleration phases. The peak of illnesses occurs at the end of the acceleration phase, which is followed by a deceleration phase, during which there is a decrease in illnesses.

Different countries or parts of countries can be in different phases of the pandemic at any point in time and different parts of the same country can also be in different phases of a pandemic.

Risk Assessment

Risk depends on characteristics of the virus, including how well it spreads among people; the severity of resulting illness; and the medical or other measures available to control the impact of the virus (for example, vaccines or medications that can treat the illness). There are no vaccines or medications specifically intended to fight this virus.

In the absence of vaccine or treatment medications, nonpharmaceutical interventions become the most important response strategy. These are community interventions that can reduce the impact of disease.

The risk from COVID-19 to Americans can be broken down into risk of exposure versus risk of serious illness and death.

Risk of exposure:

Cases of COVID-19 and instances of community spread are being reported in all states.

People in places where ongoing community spread of the virus that causes COVID-19 has been reported are at elevated risk of exposure, with the level of risk dependent on the location.

Healthcare workers actively caring for patients with COVID-19 are at an elevated risk of exposure. Close contacts of persons with COVID-19 also are at elevated risk of exposure.

Travelers returning from affected international locations where community spread is occurring also are at elevated risk of exposure, with level of risk dependent on where they traveled.

Risk of Severe Illness:
Based on currently available information and clinical expertise, older adults and people of any age who have serious underlying medical conditions might be at higher risk for severe illness from COVID-19. Based on what we know now, those at higher risk for severe illness from COVID-19 are:

- People 65 years and older
- People who live in a nursing home or long-term care facility
- People of all ages with underlying medical conditions

Chapter 9 The Remedy Across the World!

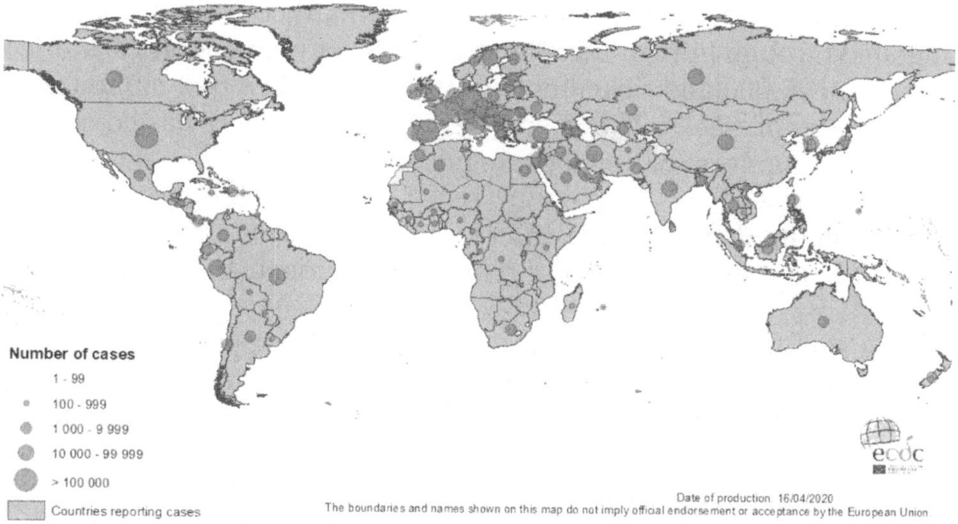

Why did it happen? Why was the world response so inept? Why did so many people die? Has washing hands and avoiding neighbors ever solved a pandemic in our past?

The whole world did not completely go in total lockdown but from those of us living through this nightmare, no matter in which country we live, it sure seems total.

As of today, there are hundreds of countries and territories that are affected by the coronavirus. It is worthy of note that Sub-Sahara Africa's numbers of cases and deaths are substantially less than most countries which have the virus.

Why? Because the areas discussed are mosquito malaria prone. The people who live there, if they are living, after DDT was banned have already been treated with chloroquine or hydroxychloroquine for years to prevent and to cure their malaria in this area of the world. For many years US travelers have taken prophylactic shots (small dosages with effects lasting several or more weeks) to not get malaria while they hunted big game in Africa. If the big game had its way, it would have banned both chloroquine and hydroxychloroquine. Meanwhile, since the evidence of a COVID-19 cure is only anecdotal, the rest of the world is struggling more so and dying with the pandemic.

The story of this nasty bug as noted in prior chapters shows that it first surfaced in the Chinese seafood and poultry market late last year, and migrated through human transfer to the 208 assorted countries killing more than 90,000 and sickening more than one and a half million in just a matter of weeks. The World Health Organization wasted no time declaring the situation a pandemic.

As a result of the penetration and spread of the virus, almost all countries instituted a number of remedies to mitigate the virus.

Almost simultaneously, countries adopted similar approaches to keep from contracting this deadly virus. Still the officials suggest that limiting face-to-face contact with others is the best way to reduce the spread of coronavirus disease 2019 aka (COVID-19). Besides quarantine, a major recommendation is to wash hands frequently with soap and water or hand sanitizers and practice as technique that has been dubbed as social distancing.

Social distancing is also called "physical distancing." It is a self-described term in many ways meaning *keeping space between yourself and other people outside of your home.* To practice social or physical distancing, here are some basic rules:

- Stay at least 6 feet (2 meters) from other people
- Do not gather in groups
- Stay out of crowded places and avoid mass gatherings

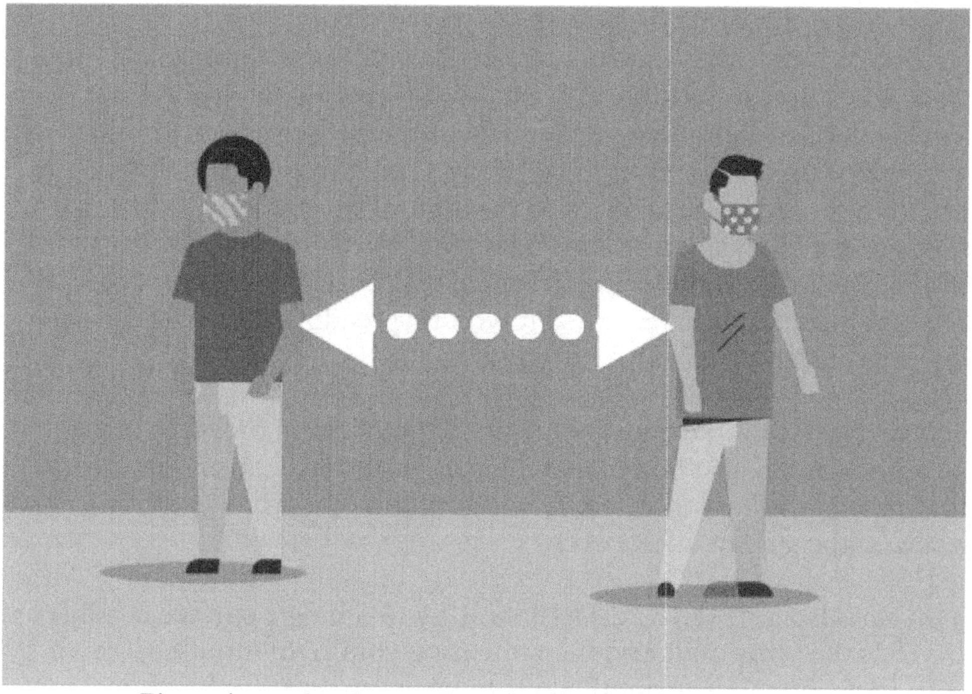

Picture is worth a thousand words What is social distancing?

In addition to everyday steps to prevent COVID-19, keeping space between you and others is one of the best tools we have to avoid being exposed to this virus and slowing its spread locally and across the country and world. Since there are no recommended therapeutics and there is no vaccine for the COVID-19, the objective is to not get it. Stay away from other humans who do not live in your immediate residence.

Here are some other recommendations:

When COVID-19 is spreading in your area, everyone should limit close contact with individuals outside your household in indoor and outdoor spaces. Since people can spread the virus before they know they are sick, it is important to stay away from others when possible, even if you have no symptoms. Social distancing is especially important for people who are at higher risk of getting very sick – those with other health issues and those over 65-years of age.

Why should I practice social distancing?

COVID-19 spreads mainly among people who are in close contact (within about 6 feet) for a prolonged period. Spread happens when an infected person coughs, sneezes, or talks, and droplets from their mouth or nose are launched into the air and land in the mouths or noses of people nearby. Just because you don't see it happening, does not mean it is not happening. Imagine that everybody you speak with or come closer than 6 feet or so to you is spreading these contagious droplets and this makes you even safer.

The droplets can infect just by being in the opening passages of your airways but can also be inhaled into the lungs. Recent studies indicate that people who are infected but do not have symptoms likely also play a role in the spread of COVID-19.

You can also get the COVID-19 virus by touching a surface or object that has the virus on it and then touching your own mouth, nose, or eyes. However, this is not thought to be the main way the virus spreads. COVID-19 can live for hours or days on a surface, depending on factors such as sun light and humidity. Social distancing helps limit contact with infected people and contaminated surfaces.

Although the risk of severe illness may be different for everyone, anyone can get and spread COVID-19. Everyone has a role to play in slowing the spread and protecting themselves, their family, and their community. For some people, the risk of not paying attention to social distancing can be to suffer death after a long illness with the virus.

Here are some tips for social distancing

- Follow guidance from authorities where you live.
- If you need to shop for food or medicine at the grocery store or pharmacy, stay at least 6 feet away from others.
- Use mail-order for medications, if possible.
- Consider a grocery delivery service.
- Cover your mouth and nose with a cloth face cover when around others, including when you have to go out in public, for example to the grocery store.

- Stay at least 6 feet between yourself and others, even when you wear a face covering.
- Avoid large and small gatherings in private places and public spaces, such a friend's house, parks, restaurants, shops, or any other place. This advice applies to people of any age, including teens and younger adults. Children should not have in-person playdates while school is out. To help maintain social connections while social distancing, learn tips to keep children healthy while school's out.
- Work from home when possible.
- If possible, avoid using any kind of public transportation, ridesharing, or taxis.
- If you are a student or parent, talk to your school about options for digital/distance learning.

Stay connected while staying away.

It is very important to stay in touch with friends and family that don't live in your home. Call, video chat, or stay connected using social media. Everyone reacts differently to stressful situations and having to socially distance yourself from someone you love can be difficult. Read tips for stress and coping.

What is the difference between quarantine and isolation?

We have heard there is some concern out there about the differences between isolation and quarantine. The following should help in their differentiation:

Quarantine

Quarantine is used to **keep someone who** might **have been exposed to COVID-19 away from others**. Someone in self-quarantine decide on their own to stay separated from others, and they limit movement outside of their home or current place. A person may have been exposed to the virus without knowing it (for example, when traveling or out in the community), or they could have the virus without feeling symptoms. Quarantine helps limit further spread of COVID-19.

Isolation

Isolation is used to **separate sick people from healthy people**. People who are in isolation should stay home. In the home, anyone sick should separate themselves from others by staying in a specific "sick" bedroom or space and using a different bathroom (if possible).

What should I do if I might have been exposed? If I feel sick? Or have confirmed COVID-19?

All people should self-monitor all the time.

When monitoring yourself, be alert for symptoms; do the following

- Watch for fever,* cough, or shortness of breath.
- Take your temperature if symptoms develop.
- Practice social distancing. Maintain 6 feet of distance from others, and stay out of crowded places.
- Contact a physician if you believe you have symptoms

What should you do if you feel healthy but: you recently had close contact with a person with COVID-19, or you recently traveled from somewhere outside the U.S. or you just came from a cruise ship or river boat

- Self-Monitor
- Self-Quarantine
- Check your temperature twice a day and watch for symptoms.
- Stay home for 14 days and self-monitor.
- If possible, stay away from people who are high-risk for getting very sick from COVID-19.

If you feel healthy but have been diagnosed with COVID-19, or Are waiting for test results, or have symptoms such as cough, fever, or shortness of breath

- Self-Isolate
- Stay in a specific sick room or area and away from other people.

- If possible, use a separate bathroom.
- Read as much important information about caring for yourself or someone else who is sick.

Look up such important information on the Internet or ask your family doctor about:

- How to Protect Yourself
    - Social Distancing
    - Face Mask
- Cleaning and Disinfecting Your Home
    - Lysol
    - Wash hands with soap & water
- Gatherings and Community Events
    - Social Distancing
    - Masks

Chapter 10 How to Resume Your Life

Times Square Establishment—A SIGN of the times

Should we stay closed and separated?

Several months ago, well before I wrote this chapter, the New York Times Magazine wrote a piece about restarting America. After a read of their article one would not necessarily conclude that we should ever restart when government is being so helpful to the people right now.

Here is how they began their article:

"The politics of the coronavirus have made it seem indecent to talk about the future. As President Trump has flirted with reopening America quickly — saying in late March that he'd like to see "packed churches" on Easter and returning to the theme days ago with "we cannot let this continue" — public-health experts have felt compelled to call out the dangers. Many Americans have responded by rejecting as monstrous the whole idea of any trade-off between saving lives and saving the economy. And in the near term, it's true that those two goals align: For the sake of both, it's imperative to keep businesses shuttered and people in their homes as much as possible.

In the longer run, though, it's important to acknowledge that a trade-off will emerge — and become more urgent in the coming months, as the economy slides deeper into recession. The staggering toll of unemployment has reached more than 22 million in just the last four weeks. There will be difficult compromises between doing everything possible to save lives from COVID-19 and preventing other life-threatening, or -altering, harms.

When can the US ethically bring people back to work and school and begin to resume the usual rhythms of American life? The times brought five people together in a video-conference and asked them point blank, what would it take to reopen America. Of course there is the current battle of who has the authority—the President or the fifty Governors. But, assuming that one will be solved, there is a lot of disagreement on when we should open. For eight governors, in fact, who have not issued stay-at home orders, the opening question is moot. They have not shut down.

If you want to see what the Times says about the subject, feel free to go to https://www.nytimes.com/2020/04/10/magazine/coronavirus-economy-debate.html. The panelists in their piece do not reflect my views on the subject.. Bioethicist Ezekiel Emanuel adopted the "Whole Lives" program for Obamacare. The "Death Panels" were part of his work in which if you were under ten or above fifty years old, Emanuel would serve you your health-care last.

He would not be on my panel. He is not pro-American in my opinion. His take is to keep America shut down until there is a vaccine, which he believes will be here in perhaps eighteen months. By then folks, if we stayed shut down with enhanced social distancing as in Michigan, China would own America. Therefore, I am not interested at all in the Emanuel Plan.

What if you knew that the number of deaths for the flu this year were more than four times the number of deaths from the coronavirus. It is by the way. In the U.S. alone, the flu this year (also called influenza) has caused an estimated 38 million illnesses, 390,000 hospitalizations and up to 62,000 deaths this season. Think about that. It's like that every year and we do not close down the country. COVID-19 has caused about 29,000 deaths in the US this year because we have no vaccine and won't accept the best therapeutic as our gold standard—hydroxychloroquine. The # of deaths would be significantly less if hydroxychloroquine were adopted as both a prophylaxis and as a possible cure.

Republican Governors Split on Trump Stay-at-Home Skepticism

By

Amanda Albright

and

Jonathan Levin

March 25, 2020, 2:54 PM EDT *Updated on March 25, 2020, 5:14 PM EDT*

6:42

President Donald Trump is a realist about public-health policies that could harm the economy and the American public. At this point of the pandemic, he sees the risk of economic collapse as a major cause for loss of life in the future—an not because of the flu or a virus. Trump has attracted crucial allies among Republican leaders overseeing the movement of millions of Americans. People are sick of staying at home waiting for their businesses be destroyed. They want to fight.

Florida Governor Ron DeSantis has resisted enacting orders that would restrict people to their homes. His philosophy is unlike what other large states such as New York and California have done. Instead, his office has advised that people older than 65 should stay inside. Mississippi Governor Tate Reeves also balked, saying the state isn't China and the US has a Constitution. Alabama Governor Kay Ivey has said she has no plans for a state shelter-in-place order, so the City of Birmingham with a Dem Mayor approved its own Tuesday. Democrats hate citizens having freedom.

Other Republicans, like Ohio's Mike DeWine, have been leading the charge to slow the virus despite the consequences. Economists suggest that social distancing could limit infections and prevent as many as 600,000 additional U.S. deaths. Most think DeWine is all wet. That has set off fierce criticism of those following the lead of Trump, who has repeatedly questioned the need, given the economic costs.

Democrats are more interested in opposing Trump than helping America. If they can get away with doing nothing, that will be their approach. Democrats don't like business and they liken the shutdown to how things would be in a socialist country so why change things by going back to work when that is exactly what Democrats do not want..

The billionaires on Wall Street are playing the hedge funds and other financial instruments and they are making money while the rest of the country is hurting. Regular Democrats think Wasll Street is for Republicans but it is not so. The Billionaires on the Street do not want the people to come back to work. They are making their billions without the people working.

And, so, the Wall Street types business leaders on the phone call with Trump on April 15, urged Trump to dramatically increase coronavirus testing no matter how long the country has to shut down. During first call to discuss reopening country, executives tell president more testing is needed to convince public to return to work

We will discuss that and other factors of reopening the country in this chapter.

What do you do when you don't know? Hopefully when asked, you will admit that you do not know. Unfortunately, the media in the US uses a different standard. They want to know what Trump thinks about the answer to the question. If Trump says X, the media, without any research of their own know the answer intrinsically with no further study. They know that it is not X, and so it is Y, if there are just two options.

Using media logic, several weeks after I heard Dr. Oz and some pretty credentialed epidemiologists and cardiologists touting hydroxychloroquine and even chloroquine as a game changer in the fight for a cure and a preventative (prophylaxis) for the coronavirus, I was very encouraged.

I began to study all I could (without being a doctor) to see if there was more than anecdotal evidence for the claim. Let me say that I learned that most doctors use the drug to treat themselves so they are not infected and though that is not science, it is great news. Moreover, in a recent global survey, Hydroxychloroquine was rated 'most effective therapy' by doctors for coronavirus. Yes, this drug known for treating malaria is being used extensively by U.S. doctors mostly for high-risk COVID-19 patients.

The FDA has authorized emergency use of the malaria medicines for severely ill coronavirus patients who are already in the hospital. Emergency use is a lower bar than full-on approval, which requires piles of clinical data about a drug's effectiveness and safety, pointed out Politico. Those answers are still in the works, and the FDA has acknowledged that even if the drugs are effective, doctors don't know the best dosing yet to treat the virus. However, the doctors have figured that out already. Moreover, they use it for more than just emergency us for COVID-19.

Chloroquine has been safely used since 1944 (about 65 years) as an antimalarial drug. Here is its full set of FDA approved uses

(1) For the treatment of uncomplicated malaria due to P. falciparum, P. malariae, P. ovale, and P. vivax.
(2) Chronic discoid lupus erythematosus and systemic lupus erythematosus in adults and
(3) Treatment of acute and chronic rheumatoid arthritis in adults.

The FDA recently posted information regarding worldwide shortages of hydroxychloroquine and chloroquine to its drug shortages webpage due to a significant surge in demand, notably as a result of US President Donald Trump's pronouncement that the drug could be a 'game changer' in the fight against COVID-19. This is not good news and must be addressed quickly with the same dispatch as the N95 masks.

Trump already ordered a stockpile of 30 million doses but he needs to ramp up production to fulfill the need for using the drug as a prophylaxis and as a cure for COVID-19 when America reopens. The American population is in excess of 330,000,000.

Physicians use Hydroxychloroquine worldwide

Valerie Richardson of The Washington Times reported a week ago on Thursday, April 2, 2020 that an international poll of more than 6,000 doctors released Thursday found that the antimalarial drug hydroxychloroquine was the most highly rated treatment for the novel coronavirus. For me that means I want it if I am sick with the virus or as a prophylaxis.

The survey was conducted by Sermo, a global health care polling company, of 6,227 physicians in 30 countries. They found that 37% of those treating COVID-19 patients rated hydroxychloroquine as the "most effective therapy" from a list of 15 options. These are doctors.

Of the physicians surveyed, 3,308 said they had either ordered a COVID-19 test or been involved in caring for a coronavirus patient, and 2,171 of those responded to the question asking which medications were most effective. Only when the scientific community accepts information such as this and offers real opinions about the positive aspects of its efficacy will this treatment get the full respect that it deserves from the world. Then, many real lives will be saved.

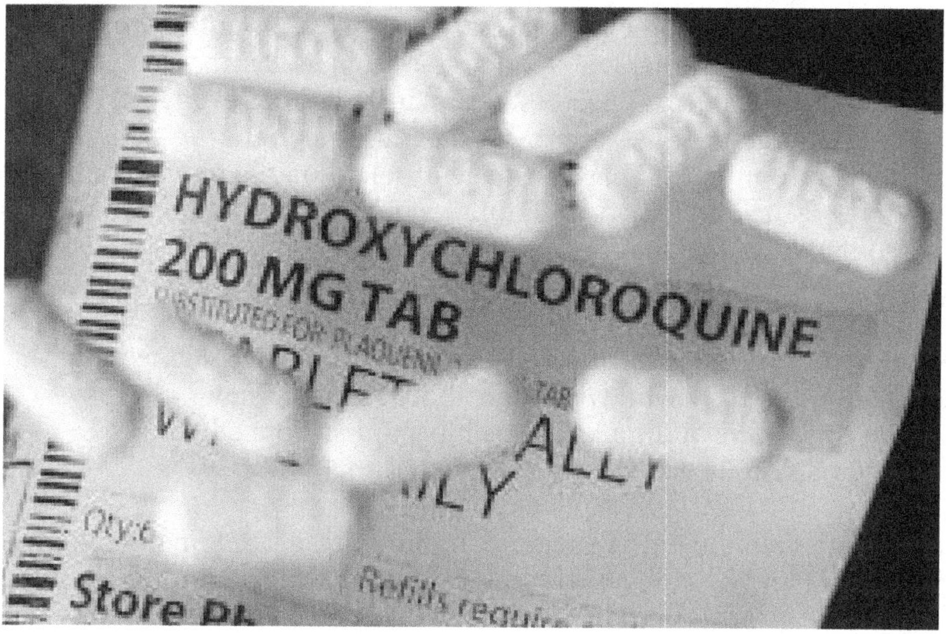

The activist press is against hydroxychloroquine after finding out that from the daily briefings that President Donald Trump authorized Peter Navarro to find thirty million or more doses from around the world from countries who had chosen not to export the medicine. Navarro found them and brought them to the stockpile. If Trump is for anything, the activist press is against it for no other reason.

President Donald Trump has expressed his interest in the merits of the anti-malarial drug hydroxychloroquine as a coronavirus game changing cure and as a prophylaxis. I learned about it from two doctors on the Laura Ingraham show and the evidence is compelling

Drs. Ramin Oskoui, cardiologist and CEO of Foxhall cardiology, and Stephen Smith, founder of the Smith Center for Infectious Diseases and Urban Health, told their story to Laura Ingraham on Fox News. The activist opposition press refute all of the evidence presented because they fear a cure for the coronavirus would assure a victory for

Trump in the November elections. It is amazing that this activist opposition media does not care what happens to America as long as they can discredit the president.

Democrat Representative cured by hydroxychloroquine

Despite the opinion of the lying press, nonetheless, former coronavirus patients like actor Daniel Dae Kim and Michigan Democratic State Representative Karen Whitsett swear by it. However, even after receiving emergency FDA approval, the anti-malarial drug hydroxychloroquine still has an image problem on the political left after being touted by President Trump. When the left found Trump favored the drug, that was enough for them to be against it—unless of course they get the virus.

Democratic State Representative Karen Whitsett from Michigan has been telling anybody who will listen about her good news. She says that the controversial drug hydroxychloroquine stopped her coronavirus symptoms "within a couple hours."

The FDA issued the emergency-use authorization late Sunday for chloroquine and its next-generation version, hydroxychloroquine, as treatments for the novel coronavirus, fueling the political back-and-forth that erupted March 19 when Mr. Trump called hydroxychloroquine a potential "game changer."

There is enough evidence for me that the drug is safe and effective as a cure and as a prophylaxis that I wrote two articles for my local paper. I was called the other day telling me that the first was about to be printed in the Citizens Voice.

With all of the fuss for the first month of the shutdown with hospitals and EMTs having mask shortages and doctors and nurses having to go to work without the proper PPE's (personal protective equipment, I penned this article

Here is the Short Letter in italics that I sent to the editor.

Date: Wed, 25 Mar 2020 19:11:59 -0400
From: "Brian W. Kelly"

Subject: What if we no longer needed N95 masks?

I have no medical credentials to be considered but I offer this anyway.. My favorite poem by Emily Dickinson starts with I'm nobody, who are you? I know who I am.

In my whole career with IBM I was a problem solver. Problems that we did not know about yesterday, we had to solve today. When I get a problem in my head and possible solutions, I did not stop until my solution was proven wrong and I had to move on to possible solution #2, or 3, or whatever.

I am intrigued by the studied capabilities of hydroxychloroquine, an effective malaria drug to fight COVID-19. It is in clinical tests in NY since Tuesday but I hear nothing about it. In the French study that has been popularized by Dr. Oz, we know that the French have cured all members of their small study. It took one person a few days longer to show relaxed symptoms. But that person too was cured. Hydroxychloroquine is also seen as a great prophylaxis (preventative). It is used so that those heading to malaria infested countries do not contract malaria.

It is said anecdotally that it does the same for the coronavirus. In other words, it can prevent one from contracting the virus.

What if it works? Does anybody know? It seems like nobody at this point cares about its preventative abilities.

Think of it?

Until the major innovations in making N95 masks for hospital workers were developed, there was and there still may be a mask shortage.

What if hydroxychloroquine were such an effective prophylaxis for the virus that anybody who took the proper preventative dosage became immune from getting the virus?

What if?

How are problems solved but first with a thesis?

What if?

The N95 mask shortage would no longer matter if a simple pill or injection could prevent the virus from affecting hospital and other EMT workers.

Should we know more about this?

We could concentrate on respirators and testing!

--- End of letter ---

After a week, I added to the facts in the letter and sent it again

2nd letter to the editor about hydroxychloroquine

Date: Wed, 8 Apr 2020 07:11:34 -0400
From: "Brian W. Kelly"
Subject: Save lives with this new therapeutic remedy

Doctors themselves are taking chloroqine and hydroxychloroquine as a cure and as a prophylactic. The people should know about this.

Additionally, with the significant # of health care workers and EMS who have caught the virus in their work it would also help in keeping them safe.

Evidence that these drugs serve as an effective prophylactic is widespread. For 65 years, chloroquine has been used in Africa and various countries to fight malaria. Where the drug is used regularly, there are no cases of covid-19.

Additionally on a recent talk show two doctors discussed

Hydroxychloroquine, Azithromycin and zinc, in a dose they said costs about $20.00 can be used as a therapeutic cure for those moderately and even severely infected. The sooner one takes the medicine however, the better. Additionally, these doctors treat Lupus patients and Rheumatoid Arthritis Patients with the hydroxychloroquine and have been doing so for years without any patient going to the hospital because of the medicine.

The drug is safe. Additionally, the one doctor treats 2000 patients for Lupus and none have contracted COVID-19. He cited another study of 14,500 Lupus patients and none have COVID. This attests to its prophylaxis capabilities.

Locally on the Frank Andrews show yesterday, Louis called in to tell his story of contracting covid-19. He said he thought he would have no issue because he was young and strong and in good health but after ten days, he had not improved and began having breathing difficulties. After a few days he called 911 and was admitted to treat the coronavirus covid-19.

He was given breathing assistance with a ventilator and he thought he might not make it. Then, he was given Hydroxychloroquine, zinc, and Azithromycin. After a day or so, he was taken off breathing assist and put on oxygen, then a reduced oxygen flow and after a few days he was discharged free of the virus. On death's door to on the way home. That is not anecdotal if you are Louis, that is very real.

A state Democrat representative from Michigan said she heard Donald Trump talking about the possibilities of the drug on TV in one of the updates. She had the virus and was getting worse. She asked her doctor if she could be put on the drug and he did so. In four hours, she told the audience she had no symptoms. Her Doctor was on the air with her.

I am thinking about when the US goes back to work. All of the serology testing for antibodies to see who is immune we know will take forever. America recently brought in 329,000,000 doses of hydroxychloroquine into its stockpile and we are making more. The threat of death is reduced substantially by this drug. When we go back to work, if doctors and the government advocate a prophylaxis of hydroxychloroquine or a cure in case of infection, all of America can go back to work with minimal risk. Don't you think the people in your circulation area of northeastern PA should know about this. Please.

Hospital workers, EMS/EMT workers, Police, Fire. warehouse workers, cashiers, manufacturing workers, etc. -- everybody can be safe rather than sorry.

-- End of 2nd email --

My two favorite antiviral drugs

1. Hydroxychloroquine
2. Remdesivir

An existing, easy-to-produce medicine that proved effective at treating or preventing SARS-CoV-2 infection would provide the fastest relief for patients and doctors. As noted, the early hope is on hydroxychloroquine and chloroquine, and many hospitals, including the University of California, San Francisco, and the University of Washington, include them in their treatment guidelines. This anti-viral is approved for uses but not as a general treatment for COVID-19. My opinion is that it should have full FDA approval but I am not a doctor. According to the reports, it has saved lives.

1. Hydroxychloroquine, aka **Plaquenil**

Some doctors are combining hydroxychloroquine with azithromycin, an antibiotic. Much of the published evidence comes from a very small French study and reports from China. Larger, more rigorous clinical trials are starting, but they will take time. Favipiravir, a flu drug shown in Japan, appeared beneficial in another small study. These medicines, especially the malaria drugs, which are being mass-produced, will be used by doctors on the front lines, but we will have to wait for evidence of whether they are benefitting patients and how much.

2. Remdesivir

Timeline: First data could come in April.

Remdesivir, an antiviral medicine that failed as an Ebola treatment, was initially developed to work against a different coronavirus. There's some evidence that it benefits Covid-19 patients. Its maker, Gilead, has been working with researchers and governments around the world to get clinical trials up and running.

The company has said to expect results in April. Six large studies are in progress, with the first, in severely ill patients in China. It was due to finish as early as April 3, according to a government website.

A study in patients with milder disease will also finish in April, with two more due in May. In the meantime, Gilead has made the drug available to hundreds of patients on a compassionate use basis.

However, it recently said that, due to overwhelming demand, it would suspend access to the drug for all but pregnant women and children as it works to create a more systematic way of giving it out without interfering with clinical trials. This new system should be in place soon. Remdesivir must be given intravenously.

University of Minnesota tests all-in for hydroxychloroquine

Each day, other than supply, it looks better for hydroxychloroquine. It is getting lots of attention by lots of epidemiologists and the University of Minnesota.

At least three clinical trials for hydroxychloroquine are trying to establish whether the decades-old malaria medication can prevent COVID-19 infections in frontline health-care workers as hospitals across the country scramble to secure enough gowns and masks for their employees.

This includes two clinical trials at the University of Minnesota testing hydroxychloroquine in health care workers reporting pre- and post-exposure to the novel coronavirus. A third trial, funded by a government agency, wants to know if the drug can prevent infections in 15,000 health care workers. See my letter to the editor earlier in this section.

There is growing concern that the current strain on the health care system and its workers isn't sustainable given the high rates of exposure faced by clinicians working in frontline emergency rooms, intensive care units, and newly established COVID-19 units. At the same time, clinicians are being asked to wear one mask per shift or reuse them at some hospitals.

The COVID-19 pandemic has sickened more than 1 million people worldwide, including over 600,000 in the U.S. Over 32,000 people have died in the US. In the U.S., more and more clinicians are contracting the virus, and some are dying.

"The lack of workplace and patient safety right now is catastrophic," Rebecca Givan, an associate professor of labor studies and employment relations at Rutgers University, said in an email.

"Hospitals need to be honest with their workers, and do everything in their power to keep workers safe so that they can continue to provide desperately needed patient care without jeopardizing their own health or that of their families." Hydroxychloroquine can save their employee's lives.

Serological testing for antibodies

A week ago when Trump announced that Peter Navarro had procured over 29 million doses of hydroxychloroquine for the storehouse, the drug began to surge out of the storehouses and into pharmacies across the country. Now, just like there is a shortage of COVID-10 testing and serological testing supplies and the ability to get readouts, there will soon be a shortage of this powerful anti-viral drug and so the stockpile needs to be replaced. It is my opinion that hydroxychloroquine is vital to the success of an American economic restart.

This note is from the Commissioner of Food and Drugs, Food and Drug Administration, Stephen M. Hahn M.D. It explains the tests for antibodies which many beleive will play a role in reopening America. Here is the commissioner's note:

> Serological tests measure the amount of antibodies or proteins present in the blood when the body is responding to a specific infection, like COVID-19. In other words, the test detects the body's immune response to the infection caused by the virus rather than detecting the virus itself. In the early days of an infection when the body's immune response is still building, antibodies may not be detected. This limits the test's effectiveness for diagnosing COVID-19 and why it should not be used as the sole basis to diagnose COVID-19.
>
> Serological tests can play a critical role in the fight against COVID-19 by helping healthcare professionals to identify individuals who have overcome an infection in the past and have developed an immune response. In the future, this may potentially be used to help determine, together with other clinical data, that such individuals are no longer susceptible to infection and can return to work. In addition, these test results can aid in determining who may donate a part of their blood called convalescent plasma, which may serve as a possible treatment for those who are seriously ill from COVID-19. This is why Vice President Mike Pence called on the laboratory community to develop serological tests for COVID-19.

In March, the FDA issued a policy to allow developers of certain serological tests to begin to market or use their tests once they have performed the appropriate evaluation to determine that their tests are accurate and reliable. This includes allowing developers to market their tests without prior FDA review if certain conditions outlined in the guidance document are met. The FDA issued this policy to allow early patient access to certain serological tests with the understanding that the FDA has not reviewed and authorized them.

The FDA can also authorize tests for COVID-19 under an Emergency Use Authorization (EUA). To date, FDA has authorized one EUA for a serological test that is intended for use by clinical laboratories.
Since the FDA issued the policy, over 70 test developers have notified the agency that they have serological tests available for use. However, some firms are falsely claiming that their serological tests are FDA approved or authorized, or falsely claiming that they can diagnose COVID-19. The FDA will take appropriate action against firms making false claims or marketing tests that are not accurate and reliable.

The FDA, an agency within the U.S. Department of Health and Human Services, protects the public health by assuring the safety, effectiveness, and security of human and veterinary drugs, vaccines and other biological products for human use, and medical devices. The agency also is responsible for the safety and security of our nation's food supply, cosmetics, dietary supplements, products that give off electronic radiation, and for regulating tobacco products.

How do we open up America?

One of the approaches to opening up America has to do with seeing who has had the virus in the past so that they can gain a certification to prove that person cannot be infected again. If there were an adequate supply of the antibody tests, this would be difficult to implement on a good day.

The notion of having to prove you're OK to get your freedom to work and walk around certificate may not fly well in a free country. That can be an obstacle as the day gets closer that we open up the country. I think we need a simpler approach. Serology testing may make us feel that we are OK but proving it to an employer or anybody else is going to be problematic. There are after all, 330 million citizens in just America to worry about. There are 7.8 billion worldwide.

There was a headline today that certainly got my attention. It said:

DON'T COUNT ON ANTIBODY TESTS TO REOPEN AMERICA

It offered that blood tests that measure a person's antibodies to the coronavirus could be a powerful tool to determine when it's safe to reopen the country. That is correct.

But after all the good news about the creation and development of such tests, concerns now exist about the accuracy and availability of the tests. Here is how the tests work:

Like the FDA Commissioner said: They detect whether a person has ever been exposed to the virus. But, there are many different tests than the single FDA approved test. These tests are different from those used to diagnose the disease. There are those who believe the existence of all these tests could hamper plans to allow Americans back to work and school.

Entrepreneurial America has created more than 90 different antibody tests. They are all now on the market, but only one has been authorized by the Food and Drug Administration. Why is this? The others "may not be as accurate as we'd like," agency FDA chief Stephen Hahn said recently as talk is ramping up about lifting the shutdown. Hahn's FDA has not verified the other 90 tests as being effective. Surely John Q. Public is not qualified to make this determination.

Public health experts are now warning that just because a person has antibodies to the coronavirus does not necessarily mean that they are immune to the virus, according to David Lim ...... But the antibody testing push comes as President Donald Trump is laser-focused on reopening the economy and governors on both coasts work on plans for a regional restart.

Rhode Island Gov. Gina Raimondo said her administration is already conducting a "deep dive, industry by industry" for guidelines to a "new normal,"

My personal concern is that government can easily guess wrong on who can do what and when in any piecemeal approach to relieving the shutdown. Unlike Dr. Fauci, I advocate turning the switch so we do not have to designate a ***grand determinator*** at the federal or state level to determine what are the gauntlet points that must be accomplished for a person to be declared free of disease, and then what? ?

So, what do we do about new screening and training for businesses that reopen. What must employees do to qualify? Where do they go? Pennsylvania for example has no testing facilities as of today—no drive throughs. Governor Raimondo is one of six northeast governors working together in a new working group announced Monday. There is a lot of hope for this group but there are a lot of pitfalls. Who makes the decisions for the group if they are independent of the president.

"Everyone is very anxious to get out of the house, get back to work, get the economy moving. Everyone agrees with that," said New York Gov. Andrew Cuomo. "What the art form is going to be here is doing that smartly, and doing that productively, and doing that in a coordinated way."

My perspective is to not put a whole load of *gotchas* and *have-tos* and , on the public or businesses. I say" "Let it happen naturally. "The people such as myself and my wife are going to be cautious entering this new open world. Let it up to us make those decisions as to what, and how. The government should just say when!

We got into this mess because it was unexpected. Nobody knew the risks of droplets or dirty hands or close contact. We have seen the case counts, and the deaths. Only a fool would act haphazardly if the switch were turned back on for the economy to start a of a certain date.

The governors' announcement came yesterday as the President asserted that he, not they, would decide when stay-at-home orders could be lifted. It also coincided with news from California Gov. Gavin Newsom, Oregon Gov. Kate Brown and Washington Gov. Jay Inslee that they are working on their own "shared approach" to restarting the West Coast economy and it does not depend on what the East Coast does.

Trump, as expected was asked about the governors' efforts during the Monday Task Force press briefing. Trump was emphatic: "a president's authority is ***total***." He added, "And that's the way it's got to be... And the governors know that." Legal scholars say the federal government lacks the power to directly order states to reopen their economies. The last thing America needs is a Constitutional fight before we can open America.

The simplest formula to open the country

Today is August 20, 2020. If I were in charge of the reopening of America, this would be the recommended approach:

It is based on every person's having a desire to survive.

1. All employees preparing to go back to work or already working or those people planning to not stay in 100% lockdown should see their doctor or clinician first; discuss their plans; schedule a test, and if you do not have antibodies, get a prescription for the hydroxychloroquine prophylaxis.

2. States need to designate areas for serology and disease testing and medicine dispensing. I know that I would not know where to go if the tests were ready tomorrow. Hydroxychloroquine prophylaxis needs to be made available as the prime solution. Patients should be able to get their dosage at:

A. Doctors' Offices
B. Clinics
C. Private Urgent Care facilities
D. Additional Facilities such as pharmacies and private areas staffed by PAs and / or nurses.
E. Doctor . Med professional recommended prophylaxis should be taken by all workers who feel the need to wear masks.

3. Shutdown is lifted across the country on May 4 to May 11, 2020

4. Behavioral recommendations (not mandatory) are continued Social distancing , hand washing etc. Avoid contact if not necessary. A new rule book of how to stay safe should be put together and made available

5. All businesses, entertainment, including restaurants, plays, movies, gymnasiums, swimming pools, parks, etc. may open. People need to voluntarily exercise caution like during the shutdown to avoid crowds if possible. If a place is crowded, go someplace else. If you are sick, stay home

6. Mass transportation reopens. Transit workers take prophylaxis.

7. Face masks recommended until further notice but not mandatory

8. People should stay at home except when they decide to go out to movies or dinner, etc. No restrictions on visiting neighbors or family other than in hospitals and nursing homes.

9. Employees will be called back to work by employers. They have five days to report to work. Those fearful to return to work may request up to 30 days additional leave. Unemployment compensation available for the thirty days for those called back to work – perhaps at half rate.

Summary

Instead of the government determining what is best for the people, the people have already suffered for more than four weeks with the country in lockdown, and the people understand the risks by now. The people are naturally cautious and restrictions will not make us more cautious.

Over time, we may choose to brave the outside world when we feel the time is right. Plus, we may opt to dine inside a restaurant with or without a prophylaxis such as hydroxychloroquine. The medicine which would serve as a crutch so that we would not be infected. Use similar cautions to now to protect yourself from infection.

I would recommend taking the proper amount of a prophylaxis such as hydroxychloroquine to ward off the virus. This medicine may last about three weeks or perhaps longer. Renew the prophylaxis regimen as required. This would be more effective than the antibodies testing and the drug can be made more available than the test. The idea is that eventually the coronavirus will be gone.

If you think you have contracted the virus or are concerned about it, you should get tested wherever you can. If positive, ask the physician or attendant for a prescription for hydroxychloroquine with a packet of Zinc and another prescription for Azythromycin. This "three pack" has been effective in curing the virus in some people. It can cure the virus but it is not 100% guaranteed.

Go home and quarantine for 14 days while taking the medicine until the doctor says you no longer have the virus. If you do not improve, call your doctor. Like all medicine, nothing is 100% but hydroxychloroquine is one that I would be looking for if I get sick.

It should be part of the nation's recommended solution for reopening.

By all means reopen the country so we have an economy ready to go when we as a country are finally ready to stop hibernating.

Other Books by Brian W. Kelly: (amazon.com, and Kindle)

The Constitution's Role in a Return to Normalcy Can the Constitution Survive?
The Constitution vs. The Virus Simultaneous attack coronavirus and US governors
One, Two, Three, Pooph!!! Reopen Country Now! Return to normalcy is just around the corner.
Reopen America Now Return to Normalcy
Enough is Enough!Re Re: Covid, We are not children. We're adults.We'll make the right decisions.
How to Write Your 1st Book & Publish it Using Amazon KDP You can do it
REMDESIVIR A Ray of Hope
When Will America Reopen for Business? This author's opinion includes voices of experts
HydroxyChloroquine: The Game Changer
Super Bowl & NFL Championship Seasons The KC Chiefs From the 1st to Super Bowl LIV
Great Coaches in Kansas City Chiefs Football First Coach era to Andy Reid Era
Great Players in Kansas City Chiefs Football From the AFL to Andy Reid Era
Reopen America Now! How to Shut-Down Corona Virus & Return to Normalcy!
Why is Everybody Moving to the Villages? You can afford a home in the Villages
CORONAVIRUS The Cause & the Cure. Many solutions—but which ones will work?
Great Moments in Kansas City Chiefs Football. From the beginning to the Andy Reid Era
How the Philadelphia Eagles Lost Its Karma. This is the one place that tells the story
Cancel All Student Debt Now! Good for America, Good for the Economy.
Social Security Screw Job!!! Scandal: Seniors Intentionally Screwed by US Government
Trump Hate They hate Trump Supporters; Trump; & God—in that order
Christmas Wings for Brian A heartwarming story of a boy whose shoulders kept growing
Merry Christmas to Wilkes-Barre 50 Ways" for Mayor George Brown to Create a Better City.
Air Force Football Championship Seasons From AF Championship to Coach Calhoun's latest team
Syracuse Football Championship Seasons beginning of SU championships; goes to Dino Babers Era
Navy Football Championship Seasons 1st Navy Championships to the Ken Niumatalolo Era
Army Football Championship Seasons Beginning of Football championships to Jeff Monken Era
Florida Gators Championship Seasons Beginning of Football through championships to Dan Mullen era
Alabama's Championship Seasons Beginning of Football past the 2017/2018 National Championship
Clemson Tigers Championship Seasons Beginning of Football to the Clemson National Championships
Penn State's Championship Seasons PSU's first championship to the James Franklin era
Notre Dame's Championship Seasons Before Knute Rockne and past Lou Holtz's 1988 undisputed title
Super Bowls & Championship Seasons: The New York Giants Many championships of the Giants.
Super Bowls & Championship Seasons: New England Patriots Many championships of the Patriots.
Super Bowls & Championship Seasons: The Pittsburgh Steelers Many championship of the Steelers
Super Bowls & Championship Seasons: The Philadelphia Eagles Many championships of the Eagles.
The Big Toxic School Wilkes-Barre Area's Tale of Corruption, Deception, Taxation & Tyranny
Great Players in New York Giants Football Begins with great players of 1925 to the Saquon Barqley era.
Great Coaches in New York Giants Football Begins with Bob Folwell 1925 and to Pat Shurmur in 2019.
Great Moments in New York Giants Football Beginning of Football to the Pat Shurmur era.
Hasta La Vista California Give California its independence.
IT's ALL OVER! Mueller: "NO COLLUSION!"—Top Dems going to jail for the hoax!
Democrat Secret for Power & Winning Elections Open borders adds millions of new Democrat Voters
Hope for Wilkes-Barre—John Q. Doe—Next Mayor of Wilkes-Barre
The John Doe Plan & WB Plan will help create a better city!
Great Moments in New England Patriots Football Second Edition
This book begins at the beginning of Football and goes to the Bill Belichick era.
The Cowardly Congress Corrupt US Congress is against America and Americans.
Great Players in Air Force Football From the beginning to the current season
Great Coaches in Air Force Football Grom the beginning to Coach Troy Calhoun
Help for Mayor George and Next Mayor of Wilkes-Barre How to vote for the next Mayor Council
Ghost of Wilkes-Barre Future: Spirit's advice for residents how to pick the next Mayor and Council
Great Players in Air Force Football: Air Force's best players of all time
Great Coaches in Air Force Football: From Coach 1 to Coach Troy Calhoun
Great Moments in Air Force Football: From day 1 to today
Great Players in Navy Football: Navy's best including Bellino & Staubach
Great Coaches in Navy Football: From Coach 1 to Coach #39 Ken Niumatalolo
Great Moments in Navy Football: From day 1 to coach Ken Niumatalolo l
No Tree! No Toys! No Toot! Heartwarming story. Christmas gone while 19 month old napped
How to End DACA, Sanctuary Cities, & Resident Illegal Aliens . best solution remove shadowsAmerica.

Government Must Stop Ripping Off Seniors' Social Security!: Hey buddy, seniors can't spare a dime?
Special Report: Solving America's Student Debt Crisis!: The only real solution to the $1.52 Trillion debt
The Winning Political Platform for America Unique winning approach to solve big problems in America.
Lou Barletta v Bob Casey for US Senate Barletta's unique approach to solve big problems in America.
John Chrin v Matt Cartwright for Congress Chrin has a unique approach to solve big problems in America.
The Cure for Hate !!! Can the cure be any worse than this disease that is crippling America?
Andrew Cuomo's Time to Go? "He Was Never that Great!"**:** Cuomo says America never that great
White People Are Bad! Bad! Bad! Whoever thought a popular slogan in 2018 *It's OK to be White!*
The Fake News Media Is Also Corrupt !!!: Fake press / media today is not worthy to be 4th Estate.
God Gave US Donald Trump? Trump was sent from God as the people's answer
Millennials Say America Was "Never That Great": Too many pleased days of political chumps not over!
It's Time for The John Q. Doe Party… Don't you think? By Elephants.
Great Players in Florida Gators Football… Tim Tebow and a ton of other great players
Great Coaches in Florida Gators Football… The best coaches in Gator history.
The Constitution by Hamilton, Jefferson, Madison, et al. The Real Constitution
The Constitution Companion. Will help you learn and understand the Constitution
Great Coaches in Clemson Football The best Clemson Coaches right to Dabo Swinney
Great Players in Clemson Football The best Clemson players in history
Winning Back America. America's been stolen and can be won back completely
The Founding of America… Great book to pick up a lot of great facts
Defeating America's Career Politicians. The scoundrels need to go.
Midnight Mass by Jack Lammers… You remember what it was like Great story
The Bike by Jack Lammers… Great heartwarming Story by Jack
Wipe Out All Student Loan Debt--Now! Watch the economy go boom!
No Free Lunch Pay Back Welfare! Why not pay it back?
Deport All Millennials Now!!! Why they deserve to be deported and/or saved
DELETE the EPA, Please! The worst decisions to hurt America
Taxation Without Representation 4th Edition Should we throw the TEA overboard again?
Four Great Political Essays by Thomas Dawson
Top Ten Political Books for 2018… Cliffnotes Version of 10 Political Books
Top Six Patriotic Books for 2018… Cliffnotes version of 6 Patriotic Boosk
Why Trump Got Elected!.. It's great to hear about a great milestone in America!
The Day the Free Press Died. Corrupt Press Lives on!
Solved (Immigration) The best solutions for 2018
Solved II (Obamacare, Social Security, Student Debt) Check it out; They're solved.
Great Moments in Pittsburgh Steelers Football... Six Super Bowls and more.
Great Players in Pittsburgh Steelers Football ,,,Chuck Noll, Bill Cowher, Mike Tomin, etc.
Great Coaches in New England Patriots Football,,, Bill Belichick the one and only plus others
Great Players in New England Patriots Football… Tom Brady, Drew Bledsoe et al.
Great Coaches in Philadelphia Eagles Football..Andy Reid, Doug Pederson & Lots more
Great Players in Philadelphia Eagles Football Great players such as Sonny Jurgenson
Great Coaches in Syracuse Football All the greats including Ben Schwartzwalder
Great Players in Syracuse Football. Highlights best players such as Jim Brown & Donovan McNabb
Millennials are People Too !!! Give US millennials help to live American Dream
Brian Kelly for the United States Senate from PA: Fresh Face for US Senate
The Candidate's Bible. Don't pray for your campaign without this bible
Rush Limbaugh's Platform for Americans… Rush will love it
Sean Hannity's Platform for Americans… Sean will love it
Donald Trump's New Platform for Americans. Make Trump unbeatable in 2020
Tariffs Are Good for America! One of the best tools a president can have
Great Coaches in Pittsburgh Steelers Football Sixteen of the best coaches ever to coach in pro football.
Great Moments in New England Patriots Football Great football moments from Boston to New England
Great Moments in Philadelphia Eagles Football. The best from the Eagles from the beginning of football.
Great Moments in Syracuse Football The great moments, coaches & players in Syracuse Football
Boost Social Security Now! Hey Buddy Can You Spare a Dime?
The Birth of American Football. From the first college game in 1869 to the last Super Bowl
Obamacare: A One-Line Repeal Congress must get this done.
A Wilkes-Barre Christmas Story A wonderful town makes Christmas all the better
A Boy, A Bike, A Train, and a Christmas Miracle A Christmas story that will melt your heart
Pay-to-Go America-First Immigration Fix
Legalizing Illegal Aliens Via Resident Visas Americans-first plan saves $Trillions. Learn how!
60 Million Illegal Aliens in America!!! A simple, America-first solution.
The Bill of Rights By Founder James Madison Refresh *your knowledge of the specific rights for all*
Great Players in Army Football Great Army Football played by great players..
Great Coaches in Army Football Army's coaches are all great.

Great Moments in Army Football Army Football at its best.
Great Moments in Florida Gators Football Gators Football from the start. This is the book.
Great Moments in Clemson Football CU Football at its best. This is the book.
Great Moments in Florida Gators Football Gators Football from the start. This is the book.
The Constitution Companion. A Guide to Reading and Comprehending the Constitution
The Constitution by Hamilton, Jefferson, & Madison – Big type and in English
PATERNO: The Dark Days After Win # 409. Sky began to fall within days of win # 409.
JoePa 409 Victories: Say No More! Winningest Division I-A football coach ever
American College Football: The Beginning From before day one football was played.
Great Coaches in Alabama Football Challenging the coaches of every other program!
Great Coaches in Penn State Football the Best Coaches in PSU's football program
Great Players in Penn State Football The best players in PSU's football program
Great Players in Notre Dame Football The best players in ND's football program
Great Coaches in Notre Dame Football The best coaches in any football program
Great Players in Alabama Football from Quarterbacks to offensive Linemen Greats!
Great Moments in Alabama Football AU Football from the start. This is the book.
Great Moments in Penn State Football PSU Football, start--games, coaches, players,
Great Moments in Notre Dame Football ND Football, start, games, coaches, players
Cross Country with the Parents A great trip from East Coast to West with the kids
Seniors, Social Security & the Minimum Wage. Things seniors need to know.
How to Write Your First Book and Publish It with CreateSpace. You too can be an author.
The US Immigration Fix--It's all in here. Finally, an answer.
I had a Dream IBM Could be #1 Again The title is self-explanatory
WineDiets.Com Presents The Wine Diet Learn how to lose weight while having fun.
Wilkes-Barre, PA; Return to Glory Wilkes-Barre City's return to glory
Geoffrey Parsons' Epoch... The Land of Fair Play Better than the original.
The Bill of Rights 4 Dummmies! This is the best book to learn about your rights.
Sol Bloom's Epoch ...Story of the Constitution The best book to learn the Constitution
America 4 Dummmies! All Americans should read to learn about this great country.
The Electoral College 4 Dummmies! How does it really work?
The All-Everything Machine Story about IBM's finest computer server.
ThankYou IBM! This book explains how IBM was beaten in the computer marketplace by neophytes

Amazon.com/author/brianwkelly
Brian W. Kelly has written 224 books including this one.
Thank you for buying this one.
Others can be found at amazon.com/author/brianwkelly

Made in the USA
Monee, IL
04 November 2026

67385046R00056